Logic Pro
Audio and Music Production

Mark Cousins

Russ Hepworth-Sawyer

ELSEVIER

AMSTERDAM • BOSTON • HEIDELBERG • LONDON • NEW YORK • OXFORD • PARIS
SAN DIEGO • SAN FRANCISCO • SINGAPORE • SYDNEY • TOKYO

Focal Press is an imprint of Elsevier

Focal Press is an imprint of Elsevier
30 Corporate Drive, Suite 400, Burlington, MA 01803, USA
The Boulevard, Langford Lane, Kidlington, Oxford, OX5 1GB, UK

Notices
Knowledge and best practice in this field are constantly changing. As new research and experience broaden our understanding, changes in research methods, professional practices, or medical treatment may become necessary.

Practitioners and researchers must always rely on their own experience and knowledge in evaluating and using any information, methods, compounds, or experiments described herein. In using such information or methods they should be mindful of their own safety and the safety of others, including parties for whom they have a professional responsibility.

To the fullest extent of the law, neither the Publisher nor the authors, contributors, or editors, assume any liability for any injury and/or damage to persons or property as a matter of products liability, negligence or otherwise, or from any use or operation of any methods, products, instructions, or ideas contained in the material herein.

Library of Congress Cataloging-in-Publication Data
Cousins, Mark, 1972–
  Logic Pro 9 : audio and music production / Mark Cousins, Russ Hepworth-Sawyer.
    p. cm.
  Includes bibliographical references and index.
  ISBN 978-0-240-52193-0 (alk. paper)
1. Logic (Computer file) 2. Digital audio editors. I. Hepworth-Sawyer, Russ. II. Title.
  ML74.4.L64C68 2010
  781.3'4536—dc22   2009052615

British Library Cataloguing-in-Publication Data
A catalogue record for this book is available from the British Library.

ISBN: 978-0-240-52193-0
Reprinted 2011

For information on all Focal Press publications
visit our website at *www.elsevierdirect.com*

11 12 13 14   5 4 3 2

Printed in China

Typeset by: diacriTech, Chennai, India

Working together to grow
libraries in developing countries

www.elsevier.com | www.bookaid.org | www.sabre.org

ELSEVIER   BOOK AID
International   Sabre Foundation

Contents

About the Authors xvii

Acknowledgments xviii

Chapter 1 The Logic Concept **3**

    1.1 Introduction 3

    1.2 A Brief History of Logic Pro 9 4

    1.3 The Logical Advantage 6

        Complete Integration with Apple Hardware and Software 7

        Exhaustive Range of Plug-Ins and Instruments 7

        Effective Combination of MIDI and Audio Editing 7

        Flexible Audio Hardware 9

Chapter 2 Logic's Interface **11**

    2.1 Introduction 11

    2.2 What Logic Can Record 11

    2.3 The Arrange Window 12

    2.4 Editor Areas – Mixer, Sample Editor, Piano Roll, Score, and Hyper Editor 13

        Mixer (Keyboard Shortcut – X) 14

        Sample Editor (Keyboard Shortcut – W) 14

        Piano Roll (Keyboard Shortcut – P) 15

        Score (Keyboard Shortcut – N) 15

        Hyper Editor (Keyboard Shortcut – Y) 16

    2.5 Media and Lists Area 17

    2.6 Inspector Area (Keyboard Shortcut – I) 19

        Region Parameters 21

        Track Parameters 21

        Arrange Channel Strip 21

Contents

2.7 Transport Bar and Toolbar   21

2.8 Tools, Local Menus, and Contextual Menus   22

2.9 Adjusting How You View the Arrangement:
     Zooming In and Out   23

Chapter 3 Getting Connected   27

3.1 Introduction   27

3.2 USB Devices   27

     USB – An Introduction   28

     USB Audio Interfaces   28

     USB Controller Keyboards   30

3.3 FireWire Devices   32

3.4 PCI Express   34

3.5 Other Audio Equipment   35

     Microphones   35

     Monitors   37

     Headphones   38

     Outboard   39

3.6 Integration into Logic – Audio Preferences   39

3.7 Hard Drives   42

     SATA and Internal Hard Drives   42

3.8 Control Surfaces   43

3.9 Distributed Audio Processing and External
     DSP Solutions   46

     Using Multiple Macs   46

3.10 What Is MIDI?   46

3.11 Digidesign Hardware Integration with Logic   49

Chapter 4 Starting a Project   55

4.1 Introduction   55

4.2 Assets and Projects   56

     Save As…   57

4.3 Working with Tracks   58

Contents

Audio	59
Software Instrument	59
External MIDI	59
4.4 An Introduction to the Audio Mixer	60
Input Monitor (I)	61
Record Arm (R)	63
Mute (M)	63
Solo (S)	63
4.5 Using the Transport and Timeline	63
Bar Ruler	64
4.6 Your First Recording	65
Difference between Mix Levels and Monitoring Levels While Recording	67
4.7 Overdubs and Punching-In and -Out	67
Quickly Punching-In Material	67
Recording in Cycle Mode	67
Autopunch	68
Introducing Take Folders	69
4.8 Creating Further Tracks and Track Sorting	70
Creating Duplicate Tracks	70
Creating Tracks with the Next Instrument	71
Creating Tracks with Duplicate Settings	71
Sorting and Organizing Tracks	71
4.9 The Audio Bin and Importing	72
Importing Other Audio Files and CDs	72
4.10 Working with Apple Loops	73
Audio Files versus MIDI Files and Instruments	76
Importing Apple Loops into Your Session	77
4.11 Improving What the Artist Hears – Headphone Mixes	77
Monitor Mix	78
Cue Mix	78

Contents

4.12 Monitoring through Effects   80
     Accounting for Processing Latency   81

Chapter 5 Audio Regions and Editing   **85**

5.1 Introduction   85

5.2 The Big Picture – Forming Your Basic Arrangement   86

5.3 Using the Inspector   87

5.4 Working with Loops, Copies, and Aliases   88
     Making Multiple Copies   88
     Loops   89

5.5 Precise Region Editing   90

5.6 Snapping – Understanding Your Editing Grid   90
     Going Finer – Bypassing Snap and Nudging Regions   93

5.7 Resizing and Cutting Regions   93
     Cutting a Region with the Scissors Tool   93
     Topping-and-Tailing   95
     Using the Marquee Tool   95
     Snap to Transients   96

5.8 Fades and Crossfades   97
     Using the Crossfade Tool to Create Fades   97
     Changing the Drag Mode to Create Fades   99
     Using the Fade Parameter to Create Fades   100

5.9 Speed Fades   101

5.10 Quick Swipe Comping   101

5.11 Expanding the Take Folder and Creating
     Basic Comps   102

5.12 Time Slipping a Quick Swipe Comp   104

5.13 Flattening or Merging a Comp   105
     Unpacking a Comp   107

5.14 Flex Time Editing   108
     Why Use Flex Time?   108

5.15 Flex Time View and Flex Time Modes   108

Contents

Slicing	109
Rhythmic	109
Monophonic	109
Polyphonic	110
Tempophone	110
Speed	110
5.16 Transient Markers and Flex Markers	111
Creating Transient Markers	111
Flexing Multiple Audio Events	111
Flexing Single Notes	112
Deleting Flex Markers	114
5.17 Quantizing Audio with Flex Time	115
Groove Templates between Regions	116
5.18 Separating Regions Based on Transient Markers	117
5.19 Freezing a Flex Time Track to Save CPU	118
5.20 Editing Multiple Tracks Using Edit Groups	119
Creating Edit Groups	119
Toggling Groups	121
5.21 Drum Replacement	121
5.22 Sample Editor Principles	123
Rendered "Off-Line" Editing	123
Accuracy	125
Additional Functionality	125
5.23 Sample Editor Applications	126
Reversing Audio	126
Accurate Region Resizing	126
Marking Anchor Points	127
Cleaning Up Unwanted Noise	128
Audio to Score	128
Time and Pitch Machine	129
5.24 Inserting and Deleting Sections of Your Song	130

Contents

5.25 Working with Tempo and Loops 131

    Follow Tempo 131

    Timestretching a Region to Fit 132

    Working Out the Tempo of a Region/Loop 132

    Matching Your Project's Tempo to an Imported Loop 132

    Beat Mapping 134

5.26 Managing Your Audio Files and Regions 137

    The Audio Bin Tab versus the Audio Bin Window 137

    File Management 138

    Copy/Converting Files 139

    Viewing the Bin and Grouping Files 139

5.27 Using Folders and Hiding Tracks 141

    Packing a Folder 141

    Viewing a Folder's Contents 142

    Hide Tracks 143

Chapter 6 MIDI Sequencing and Instrument Plug-Ins 145

6.1 Introduction 145

6.2 MIDI Concepts 145

6.3 Creating Instrument Tracks 146

6.4 Instantiating Virtual Instruments and the Library Feature 147

    Browsing Instrument Settings in the Library 147

    Using Channel Strip Settings with Instruments 150

    Changing the Instrument Parameters 151

    Changing an Instruments Icon 151

    Transposition and Velocity 151

    Key Limit 152

6.5 Working with External MIDI Instruments 152

    How Do I Monitor MIDI Sound Sources? 155

    What If I've Got More Than One External MIDI Device? 157

Contents

My MIDI Tracks Are Playing Back Slightly Late,
How Do I Rectify This?   157

6.6 Making a MIDI Recording   158

Where's the Data?   158

Recording in Cycle Mode and Overdubs   158

6.7 Editing and Arranging MIDI Regions   161

Merging MIDI Regions   161

Working Without Fades and Crossfades   161

Timestretching MIDI Regions   161

6.8 Region Parameters: Quantizing and Beyond   162

Quantize   163

Q-Swing   163

Loop and Transpose   164

Velocity and Dynamics   164

Gate Time   164

6.9 The MIDI Thru Function   166

6.10 Advanced Quantization Options   166

Q-Strength   167

Q-Flam   168

Q-Range   168

6.11 Normalizing Sequence Parameters   169

6.12 MIDI Editing in Logic   169

6.13 The Piano Roll   170

6.14 Typical Editing Scenarios in the Piano Roll   172

Correcting Pitch   172

Correcting Timing   172

Duplicating a Phrase   173

Correcting Length Issues   174

Changing Velocity   174

Creating New Notes   176

Deleting/Muting Notes   177

6.15 Quantizing Inside the Piano Roll Editor   178

Contents

6.16 Working with Controller Data Using Hyper Draw   180

Editing Velocity in Hyper Draw   180

Editing Volume   181

Editing Other Controllers, Including Filter Cutoff   181

6.17 Hyper Draw in the Arrange Area   182

6.18 Going Further: The Piano Roll's Edit and Functions Menus   183

6.19 Intelligent Selection: The Edit Menu   183

Select Equal Events   186

6.20 Functions: Quick-and-Easy Note Modifications   186

Note Overlap Correction/Note Force Legato (Functions > Note Events)   186

Select Highest Notes, Select Lowest Notes (Functions > Note Events)   186

Voices to Channels (Functions > Note Events)   187

6.21 Step-Time Sequencing   188

6.22 The Hyper Editor   190

Editing and Creating MIDI Controller Data in the Hyper Editor   190

Adding New Controller Types and Hyper Sets   192

Editing Drums Using the Hyper Editor   193

6.23 Score Editor   193

6.24 Event List   193

Filtering Events   195

Modifying and Adding Events   195

Chapter 7 Creative Sound Design   **199**

7.1 Introduction   199

7.2 Logic's Synthesizers   199

ES2   200

EFM1   201

Sculpture   201

Contents

7.3	Understanding the ES2	202
	The Oscillators (A)	202
	The Filters (B)	203
	Amplifier and Effects (C)	204
	Modulation Matrix (D)	204
	LFOs and Envelopes	205
7.4	Working with Oscillators	205
7.5	Filters, Amplifiers, and Modulators	210
	Cutoff and Resonance	211
	Changing Volume Over Time	212
	Bringing It All Together: The Modulation Matrix	213
7.6	Global Parameters and Output Effects	216
7.7	EFM1 and Frequency Modulation Synthesis	216
	Modulation on the EFM1	220
	Refining Your EFM1 Patches	222
7.8	Component Modeling: Sculpture	222
7.9	Objects	223
7.10	The String	226
7.11	Waveshapers and Beyond	229
7.12	Modulation and Morphing	232
7.13	Creative Sampling	237
	The EXS24 Hierarchy: Zones, Groups, and Editors	237
7.14	The EXS24 Instrument Editor	239
7.15	Creating a New Instrument and Importing Samples	242
7.16	Changing Zone Properties	244
	Key Range	244
	Pitch	244
	Vel. Range	245
	Output	246

Contents

	Playback	246
	Sample	246
	Loop	247
7.17	Working with the EXS24's Groups	248
	Advanced Mapping Options	249
7.18	Editing EXS24 Instruments Using the Front Panel	250
	Basic Instrument Refinement: Velocity Scaling, Polyphony, and Envelopes	251
	Adding in the Filter	251
	Modulation Heaven!	252
	Saving Instrument Settings	253
7.19	Ultrabeat	254

Chapter 8 Mixing in Logic | **259**

8.1	Introduction	259
8.2	Channel Strips: Understanding Your Virtual Console	260
	Audio Channel Strip	260
	Instrument Channel Strip	260
	Aux Channel Strip	264
	Stereo Output Channel Strip	265
	MIDI Channel Strips	267
	Master Channel Strip	268
8.3	Organizing Your Mixer: What You Do and Don't See	268
	Mixer Views	269
	Viewing by Signal Path	272
	Dropping Out Sections of Your Mixer	272
8.4	Folders and the Mixer	273
8.5	Beginning a Mix	275
	Basic Part Leveling	275
	Applying Equalization and Compression	276
	Adding in Further Plug-Ins	278

Contents

8.6 Adding Send Effects   280

8.7 Combined Processing Using Aux Channels   283

8.8 Using Groups   285

8.9 Working with Channel Strip Settings   288

8.10 Automation: The Basics   289

8.11 Track-Based versus Region-Based   290

8.12 Automation Modes   292

    Write   292

    Touch   293

    Latch   293

    Read   293

8.13 Viewing and Editing Automation   295

8.14 The Automation Menu Options   300

Chapter 9 Mastering in Logic   **309**

9.1 Introduction   309

9.2 Different Approaches to Mastering   309

    Technique 1: iTunes   310

    Technique 2: Burn from Logic   310

    Technique 3: WaveBurner   311

9.3 Bounce to Disk   311

9.4 Audio Mastering in Logic   314

9.5 Editing Fades   314

9.6 Exporting and Burning   317

9.7 Mastering in WaveBurner   320

9.8 Processing and Editing   322

9.9 Dithering, Bouncing, and Burning   324

Chapter 10 Logic and Multimedia Production   **335**

10.1 Introduction   335

10.2 Managing Movies   335

    Extracting the Movie Audio Content   338

10.3	Global Tracks	339
	Markers	339
	Detect Scene Changes	341
	Working with Tempo and Signature Changes	342
	Alternatives	345
	Big Displays	346
	Spotting Audio	346
10.4	Synchronizing Logic	348
	Getting Locked	349
	Being the Master	351
10.5	Score Editing and Music Preparation	352
	Preparing Your MIDI Files	353
	Opening the Score Editor	354
	Adding Expression and Score Markings	355
	Shrinking to Fit: Score Settings and Score Sets	356
	Printing the Parts	357
10.6	Surround Sound in Logic	358
	Getting Started	359
10.7	Delivery Formats	365
	Bouncing to Stereo and Surround	365
	Exporting to DVD	368
	Exporting Video with Audio (Dubbing)	372
Chapter 11	**Optimizing Logic**	**375**
11.1	Introduction	375
11.2	Templates	375
11.3	Screensets and Windows	377
	Link Modes	381
11.4	Key Commands	382
11.5	The Environment	384
	Layers and Windows	384
	Objects	386

Contents

Cables 387

Setting Up MIDI Instruments in the Environment 389

Live MIDI with the Environment 395

External Control with the Environment 396

11.6 Input/Output Labels 400

11.7 Nodes and Distributed Audio Processing 401

Subject Index *407*

About the Authors

MARK COUSINS

Mark Cousins works as a composer, programmer, and engineer (www.cousins-saunders.co.uk), as well as being a senior writer for *Music Tech* magazine. His professional work includes composing music for some of the world's largest production music companies – including Universal Publishing Production Music – with broadcaster credits including BBC1, BBC2, ITV, Channel 4, Five, BBC World, and Sky One, among others. He has also had works performed by the Royal Philharmonic Orchestra, the East of England Orchestra, City of Prague Philharmonic Orchestra, and the Brighton Festival Chorus.

Mark has been an active contributor to *Music Tech* magazine since issue one. He has been responsible for the majority of cover features, as well as the magazine's regular Logic Pro coverage. As a senior writer, he has also had strong editorial input on the development of the magazine, helping it become one of the leading brands in its field.

RUSS HEPWORTH-SAWYER

Russ Hepworth-Sawyer is a sound engineer and producer with more than 14 years of experience of all things audio and is a member of the Music Producer's Guild, Association of Professional Recording Services, Audio Engineering Society, and is a Fellow of the Institute For Learning. Through MOTTOsound (www.mottosound.co.uk), Russ works freelance in the industry mainly as a mastering engineer, but he is also as a producer, writer, and consultant. Russ is currently an Associate Lecturer for the Open University and a part-time Lecturer for the London College of Music. He has taught extensively in higher education. He has also contributed to *Sound On Sound*.

Acknowledgments

MARK'S ACKNOWLEDGMENTS

I would like to thank my Mum and Dad for gifting me a Casio CZ-3000 for my Christmas present in 1987 and starting the whole ball rolling; my wife, Hannah, and our two kids, Josie and Fred, for their support, patience, and amusement over the years, and for keeping me company in the studio from time to time; Russ, who has been a longtime and sincere friend, but whose suggestion of writing a book together I should have politely declined! Neil Worley, for taking me on at *Music Tech* and showing me the hallowed profession of a being a writer; Adam Saunders, for putting up with my lack of musical output for several months; and finally, Polly, Deb, and Cath for looking after Fred in those final few hectic weeks!

RUSS' ACKNOWLEDGMENTS

I would like to thank my wife Jackie and son Tom for putting up with yet more of my absenteeism during the writing of this update – thank you!; my Mum, Jo ("Hi Mum"); my parents-in-law Ann and John for all their help; Mark for his guidance, knowledge, and for being the top chap he is; Max Wilson for his endless support and forcing me to move to Logic in the first place … thanks boss!; Jon Miller for the use of his animation file, wherever he is!; Iain Hodge and Peter Cook at the London College of Music; and Craig Golding, Danny Cope, and Barkley McKay at Leeds College of Music.

Both authors would like to thank Gorden Keppel from Apple for taking the time to technically check our words, and Catharine Steers and her colleagues at Focal Press for their support throughout the writing of this book.

In This Chapter

1.1 Introduction   3

1.2 A Brief History of Logic Pro 9   4

1.3 The Logical Advantage   6

Knowledgebases

Studio versus Express: The Many Flavors of Logic   5

Installing Logic Studio   7

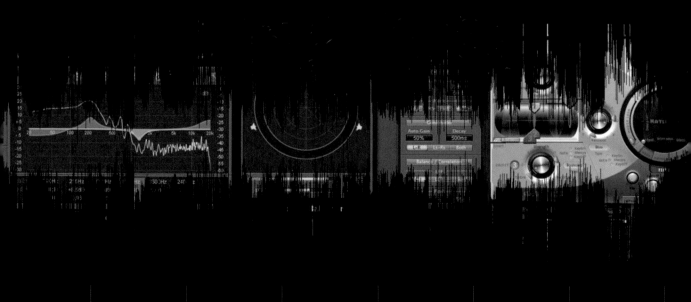

The Logic Concept

1.1 Introduction

It's hard to imagine a more complete system for music and audio production than Logic Pro 9: multitrack recording and editing, a full suite of virtual instruments and effects, and a seamless workflow that takes you from the beginning of your project right through to the delivery of the final production master. Yet, with such a complete system comes the daunting task of understanding how the elements of Logic Pro 9 knit together to produce a professional-sounding result. For example, where do you begin to start writing music or making a recording in Logic? What are the virtual instruments and plug-ins used by the professionals to create release-quality output? And how can you transform those poorly performed band recordings into a polished CD?

So, let's be clear from the start: this book isn't just another instruction manual for Logic Pro 9. Instead, we've taken a process-driven approach that appraises, understands, and explores the features of Logic Pro 9 in a way that matches the structure and order of the production process. We'll do more than just technically describe the functions of Logic Pro 9; we'll look at how the various elements of Logic Pro 9 relate to the demands of audio and music production. With all but a few exceptions, most of the chapters focus on a specific part of the production process – whether it's initial track laying, sound design, or mastering your finished mixes to produce the final CD – highlighting the relevant parts of the application that guarantee a professional-sounding audio product. We'll also look at techniques that go beyond the scope of the manual – practices like parallel compression, for example, that many engineers use and abuse on a daily basis.

If you're starting off from scratch, it's easy to be overwhelmed by the sheer size and complexity of an application like Logic Pro 9. However, it isn't essential to understand the entirety of the application to start producing music. Get to know the components that are most relevant to your way of working and build from there – use plenty of presets, Apple Loops, and so on to get you kick started – and then enjoy the process of exploring each element that little bit

DOI: 10.1016/B978-0-240-52193-0.00001-0

Figure 1.1 Logic Pro 9 includes an impressive array of features, but understanding how they integrate into the production process might not be immediately apparent.

further. Ultimately, Logic Pro 9 is a tool that will grow with your experience – a system that will surprise at every turn and open up new possibilities whenever you want to explore the software further. With this book, you'll at least have a reference to aid you in that process, but don't be afraid to experiment to find out how Logic Pro 9 best fits into your unique creative process!

1.2 A Brief History of Logic Pro 9

Like the other "old-timers" of computer-based audio production – including Cubase and Pro Tools – Logic Pro 9 is an application with a rich and long heritage in the industry. Born from the ashes of C-Lab's Notator and Creator in 1993, Notator Logic (as it was then called) was an attempt to create a visual, region-based production environment for MIDI sequencing. Building blocks, or regions of MIDI data – used to control hardware synthesizers and samplers – could be arranged on the computer screen, with a clear, visual representation of the structure of the arrangement. What was unique about Logic, though, was that the application was completely configurable – users could create virtual presentations of their studio, known as an environment, for example, or combine different editor windows in a completely configurable user interface.

Audio functionality was added to the application in 1994, with the release of version 1.7, allowing Logic users to combine both digital audio and MIDI data all in the same arrangement (although initially, only with expensive Digidesign audio hardware). Virtual instruments followed in 2000, making the system a complete production environment where a track could be composed, mixed, and mastered all in one computer, and arguably, without the need for any extra third-party software. Although revolutionary at the time, this method of production has now become the norm, with many musicians and engineers largely working entirely "in the box."

Apple acquired the company that originally developed Logic – Emagic – in 2002, with its programming team joining Apple's, and Logic Pro becoming part of Apple's prized suite of media-based applications, including Final Cut Studio and Aperture 2. The partnership led to many of Logic Pro's technologies migrating into other Apple applications – most notably with the introduction of GarageBand – as well as Apple making Logic Pro an increasingly more price-competitive option, with both the absorption of previously optional software components into the main application (like Space Designer, the EXS24 Sampler, and the ES2 synthesizer), and, with the release of Logic Studio, a halving of its retail price.

With the introduction of Logic Pro 8 and Logic's current incarnation – Logic Pro 9 – Apple has made some big moves to make the application significantly easier to use and much more in line with the usability of its other media products. As a result, it's never been a better time for new users to join the Logic Pro 9 fold – both with respect to its affordability and the significantly easier learning curve!

Knowledgebase 1 ▼

Studio versus Express: The Many Flavors of Logic

So as to match your precise production needs and available budget, Logic is available in two principal versions – Logic Studio and Logic Express. As you'd expect, Logic Studio is the more complete package, with a range of ancillary applications including MainStage 2, WaveBurner 1.6, Soundtrack Pro 3, and Compressor – as well as a wealth of sound content in the form of Apple Loops, EXS24 instruments, and so on. The main component, though, is the Logic Pro 9 application itself, which is the centerpiece of any music or audio production–based activity on the Mac. In effect, the additional applications build on Logic Pro 9's core functionality – with MainStage 2, for example, allowing you to take Logic Pro 9's instruments and effects on the road, while Soundtrack Pro 3 allows you to better integrate your work with professionals working in film and TV postproduction.

Logic Express 9, on the other hand, is a more cost-effective introduction to the world of music production in Logic. Although Logic Express 9 lacks the

(Continued)

ancillary applications and full sound content of the complete Logic Studio, it does provide a feature set almost identical to that of Logic Pro 9. On the whole, the omitted features largely relate to professional applications – using time division multiplexing (TDM)/Digidesign Audio Engine (DAE), for example, distributed audio processing, or surround sound mixing. The list of available plug-ins, so important to "in the box" audio production, is almost identical, with the possible exception of a few instruments like Sculpture and the EVP88, alongside the Space Designer and Delay Designer audio plug-ins.

In writing this book, therefore, we concentrated on the main features and processes applicable to both Logic Studio and Logic Express 9 (henceforth referred to simply as "Logic"). Where appropriate, we have referred to some of Logic Studio's extra features and components – like 5.1 surround sound mixing or WaveBurner – but in most cases, a Logic Express user will be able to achieve much of what this book details.

Figure 1.2 Logic Studio is a complete music production toolkit – with a range of applications and audio content.

Figure 1.3 Logic Express 9 omits the extra applications included in Logic Studio package, as well as presents a slimmed-down set of Apple Loops.

1.3 The Logical Advantage

Opinion and debate will always rage as to the "best" digital audio workstation, but there are a number of factors that give Logic Pro 9 the edge over alternative solutions. Certainly, if you're trying to make a decision between different audio applications – all with such a compelling range of features – it's well worth understanding some of their main overriding benefits, as well as seeing whether these align with your intended method of working.

Complete Integration with Apple Hardware and Software

Being part of Apple, you can guarantee that Logic Pro 9 will make optimal use of both Apple's computing hardware and the operating system that ties it all together. For example, where other developers might lag behind certain OS updates, Logic Pro tends to be first off the block supporting major upgrades such as Snow Leopard. On top of this, Logic Pro has always stood out from the crowd in terms of its efficient use of DSP resources, suggesting a well-coded audio engine, as well as plenty of integrated components – for example, the EXS24 sampler or Space Designer reverb – that ensure a completely optimal use of your computer's processor.

Exhaustive Range of Plug-Ins and Instruments

Logic Pro 9's integral range of instruments and effects is easily the most comprehensive set available in any off-the-shelf DAW (Digital Audio Workstation). In addition to standard studio stalwarts such as compression, reverb, and equaliza-tion, Logic Pro 9 includes a number of contemporary effects and plenty of software instruments covering everything from vintage Hammond organs to cutting-edge component-modeling synthesizers. Ultimately, with such a diverse collection of tools, you can easily produce a professional, release-quality output without having to resort to additional third-party plug-ins and effects, although of course, there's no reason why you can't add these at a later point should you wish to do so. As the old saying goes, the only limit with Logic Pro 9 is your imagination….

Effective Combination of MIDI and Audio Editing

While some applications have strengths in a particular area (like MIDI or audio), Logic Pro 9 presents an effective hybrid solution for both MIDI-based

Knowledgebase 2 ▼

Installing Logic Studio

The full version of Logic Studio is now a 47-GB installation, with a range of applications (including Logic Pro 9 itself) as well as six Jam Packs of Apple Loops and instruments. You'll also need to be running Mac OS X 10.5.7 or later, using an Intel processor, with a minimum of 1 GB of memory. If you intend to use a large amount of sample-based instruments, you might also want to consider increasing your RAM resources to 2 GB or more of physical memory.

When running the installer, you'll be presented with various options for installing Logic. If you've got a large enough hard drive (150 GB+), it's well

(Continued)

worth installing the complete package, although users with smaller drives, especially with a large amount of existing data, might want to consider a more slimmed-down installation. Arguably, the biggest use of disk resources, therefore, is the Jam Packs, Sound Effects, and Music Beds installed as part of the Logic Studio Content. Try opening up these folders in the install dialog box (click on the small arrows) and deactivating the elements you don't want to install, maybe deselecting genres that don't interest you as much. As you remove items from your installation requirements, you'll notice the Space Required indicator reducing accordingly. As a guide, try to leave at least 50% of your drive's resources still available after installation.

On the whole, the applications themselves (including Logic Pro 9, Soundtrack Pro 3, MainStage 2, and WaveBurner 1.6) don't tend to use up anywhere near as much space as the sound content, allowing even the most jam-packed laptop hard drive to accommodate Logic 9. Slimmed down to the minimum, you should be able to get the application-only installation down to around 3 GB. If you own any legacy Emagic hardware (like the Unitor/AMT8 MIDI interfaces), do check the Hardware Support folder so as to install the latest and most up-to-date drivers for these products.

Figure 1.4 The full Logic Studio installation can take up to 47 GB of drive space, although it is possible to use the custom install feature to add or remove items from your installation requirements.

composition and audio recording and editing, which probably explains why so many working composers operate completely within the realms of Logic Pro 9. Its flexibility also makes it possible to use Logic Pro 9 in a wide range of audio production environments including 5.1 and surround sound mixing, sound for film, and other multimedia applications.

Flexible Audio Hardware

Rather than being tied into one type of audio hardware, Logic Pro 9 users have the opportunity and flexibility to select the audio hardware for the way they want to work. For example, a simple laptop solution could just make use of the internal audio hardware, or simple two-in two-out USB interface, while professional users could use dedicated Apogee audio interfaces, or even use Logic Pro 9 as the front end to a full Pro Tools|HD rig.

In This Chapter

2.1 Introduction   11

2.2 What Logic Can Record   11

2.3 The Arrange Window   12

2.4 Editor Areas – Mixer, Sample Editor, Piano Roll,
     Score, and Hyper Editor   13

2.5 Media and Lists Area   17

2.6 Inspector Area (Keyboard Shortcut – I)   19

2.7 Transport Bar and Toolbar   21

2.8 Tools, Local Menus, and Contextual Menus   22

2.9 Adjusting How You View the Arrangement:
     Zooming In and Out   23

Knowledgebase

What Is a Region?   14

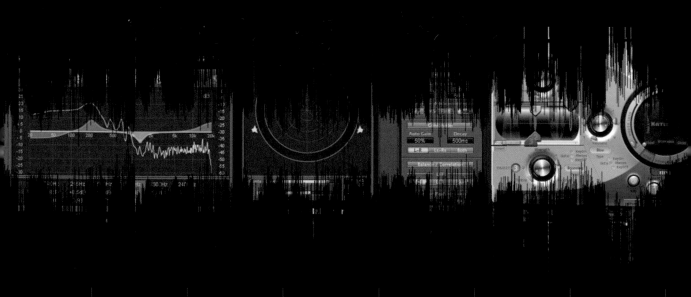

Logic's Interface

2.1 Introduction

Negotiating the interface of any audio application is vital to understanding how it works and the precise method involved in creating a finished audio recording. Logic's interface is no exception, providing its own unique slant on the production process: with a number of editor windows, mixers, media browsers, and so on, all used to sculpt and refine your audio output. Navigate Logic's interface in an informed way, therefore, and the application will become a seamless and enjoyable part of your audio production workflow, rather than a stumbling block to your creative exploits.

In this chapter, we're going to take a look at both the overarching principles of using Logic, alongside the specific components of the Logic interface and how these integrate into the production process. From this important stepping-stone, we can then begin the process of creating our own projects and taking a more detailed look at Logic's role and input in audio and music production.

2.2 What Logic Can Record

Logic principally works with two types of data – audio and MIDI. Physical instruments and external sound sources are recorded directly into Logic as audio sound files. You might, for example, plug your guitar straight into a DI input on your audio interface or set up a number of microphones connected to your interface's mic preamps – all of which will be recorded as audio files directly into Logic's Arrange window. Once recorded, you can edit and mix these files to create your finished track, ready to be burned on to CD.

As an alternative to audio files, you can also record MIDI data directly into Logic using an attached MIDI keyboard, which, in turn, can be used to control Logic's integral virtual instruments (or other third-party Audio Unit instruments), as well as external hardware MIDI synthesizers and samplers. Unlike audio recording,

DOI: 10.1016/B978-0-240-52193-0.00002-2

MIDI production provides an unprecedented amount of scope over both the performance and sound of the music, although it can sometimes lack the life and energy of music performed by real musicians.

Of course, it's highly likely that most projects in Logic will use a combination of both audio and MIDI recordings together to create the optimal presentation of your track. In that respect, the combined power and functionality of the Arrange window in handling such projects makes Logic a superb production solution.

2.3 The Arrange Window

The Arrange window is the nerve center of your work in Logic, providing a range of different editing and arrangement features to piece your project together. Placed in the middle of the interface, and covering the majority of the screen, is the Arrange area, displaying a list of current tracks and instruments residing in the project (in the track list down the left-hand side of the Arrange area), alongside the various regions that form the structure and development of your song over time. By adding or deleting tracks or moving regions around the screen, we can visually build up the structure and arrangement of the song accordingly.

Springing up from the sides of the Arrange window are a number of other functional "areas." These areas relate to activities involved in the production process and the tools of editing, arranging, and mixing audio and MIDI data. For

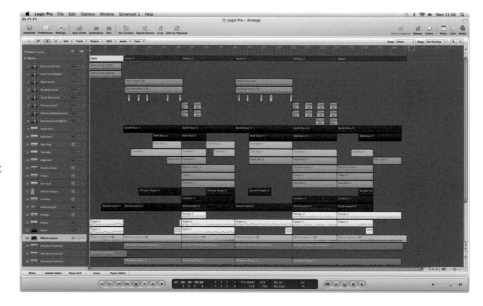

Figure 2.1 The main arrangement area provides a visual overview of the tracks used in the project, as well as the arrangement of the song defined by the various regions.

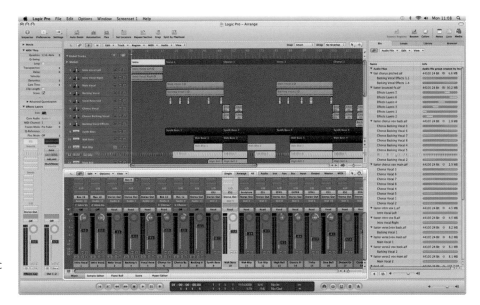

Figure 2.2 Use the other areas of the Arrange window to access more detailed application-specific editing tools and features.

example, in the initial stages of production, you might need to input the various signals to be recorded, checking their respective recording levels and making a suitable "monitor mix" – in this case, opening up the Mixer tab at the bottom of the Arrange window. Later on, you might want to import some additional audio files – using Logic's Browser – or edit the timing of a MIDI performance using the Piano Roll.

The idea with Logic's Arrangement window, therefore, is that it can provide a dynamic working environment, optimized for the tasks you need to carry out at any point in time. Although it's possible to open and shut many of these extra areas using on-screen tabs, or double-clicking on regions, it's well worth memorizing some of the important keyboard shortcuts so that you open them "on the fly." Importantly though, whatever additional area you decide to open up, you'll still have some overview of the project's arrangement, allowing you to keep a handle on how your edits affect the entirety of the track.

2.4 Editor Areas – Mixer, Sample Editor, Piano Roll, Score, and Hyper Editor

Opening from the bottom of the Arrangement window, the various editor areas within Logic allow you work with and edit your project in a number of different ways.

Knowledgebase ▼

What Is a Region?

One of the principal concepts of a DAW system like Logic Pro is the notion of the nondestructive edit. Unlike the process of editing tape – which physically cuts the tape into two – an edit in Logic simply changes the proportion of data played. In that respect, we need to make an important distinction between the "source file" or recording and a "region." In effect, a region is simply a window on the original data, playing back either a small or large part of file. You can also have a number of different regions focusing on different parts of the original file, but, ultimately, all sourced from the same material. Of course, all of these types of editing would be impossible to achieve with a physical, destructive editing medium like tape.

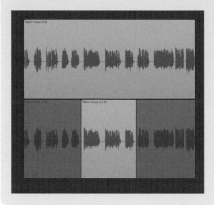

Figure 2.3 A region is simply a window on the original data, allowing you to resize the edit at any point, effectively bringing back audio or MIDI data after the original start or end point has been established.

Mixer (Keyboard Shortcut – X)

The Mixer balances the respective levels of the tracks included in your project, as well as instantiating different signal processing plug-ins (like reverb, compression, or equalization) to process each mixer channel. In addition to audio mixing, the Mixer is also the primary port of call for using virtual instruments (like synthesizers, samplers, and vintage keyboards), allowing us to create music completely within the realms of Logic. Alongside the Arrange area, the Mixer is undoubtedly one of the most important day-to-day production tools used in Logic.

Sample Editor (Keyboard Shortcut – W)

The Sample Editor provides detailed, sample-accurate audio editing of a given region in the Arrange area. Although a large amount of audio editing can be achieved in the Arrange area alone, it is the Sample Editor that really allows

Figure 2.4 The Mixer allows us to balance the mix, process tracks using audio effect plug-ins, and instantiate virtual instruments.

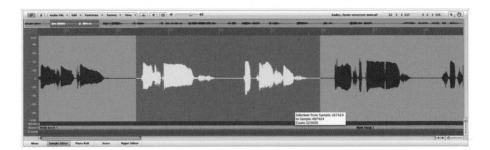

Figure 2.5 The Sample Editor allows you to edit and process your audio files in ways impossible to achieve in the Arrange area alone.

us to comfortably work with low-level details of an audio recording – precisely setting edit points, for example, or applying unique "destructive" editing techniques such as reversing, silencing, or time compression/expansion.

Piano Roll (Keyboard Shortcut – P)

The Piano Roll editor is the main MIDI editor in Logic, used to edit MIDI information that in turn controls virtual instruments or external hardware synthesizers and samplers. Using an intuitive display of note information based on the position on the piano keyboard (on the vertical axis) and the horizontal axis displaying both the note's starting point and duration, Piano Roll offers both unparalleled precision and ease-of-use for MIDI editing. Using the extended editing functions of the Piano Roll, you can also transform and edit MIDI information in ways that would be extremely time-consuming (if not impossible) to perform by hand.

Score (Keyboard Shortcut – N)

In addition to providing a useful way of visualizing and editing MIDI information for musically literate users, the Score editor is the vital conduit between your raw MIDI recordings and a finished printed score readable by real musicians!

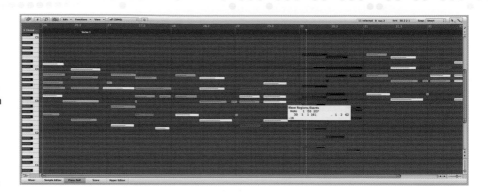

Figure 2.6 Piano Roll is Logic's main MIDI editor, with a range of editing tools and features dedicated to the precise editing of performance data.

Figure 2.7 Although slightly less versatile as a MIDI editor, the Score area is still a useful tool for turning your MIDI performances into a finished score.

Although not everybody will want to use the Score editor, it does illustrate Logic's intention to be a versatile and competent music production tool in the widest sense.

Hyper Editor (Keyboard Shortcut – Y)

This final form of MIDI editor offers a more specialized solution to working with controller data – like volume or pan – as well as offering an alternative means of programming and creating drum tracks. Information is presented within a unique series of vertical beams, allowing you to quickly draw in controller sweeps, for example, or complicated drum patterns full of sixteenth notes. As with the Score area, the Hyper Editor might not be to everyone's tastes (there are alternative ways of achieving many of its main applications), but it does form an interesting alternative for working with several different lanes of controller data outside the Piano Roll editor.

To change the respective proportion of screen devoted to the main Arrange area and the various editors, click and drag on the thin gray line between the Arrange area and bottom editor area. This allows you to optimize the view based on your current priority – for example, expanding the Mixer when you need full access to its compression and equalization settings or using a minimized Mixer (with just the main fader positions viewable) when you want to quickly rebalance the mix in relation to the arrangement.

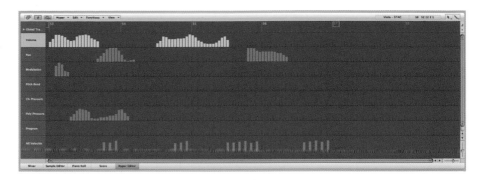

Figure 2.8 Hyper Editor displays its information as a series of unique vertical beams. As such, the Hyper Editor forms a useful solution for dealing with MIDI controller data.

Figure 2.9 Change the relative window size (using the thin gray line between the areas) to prioritize the view based on your current production activity.

2.5 Media and Lists Area

Any project in Logic will be potentially composed of a number of different audio files, plug-in settings (for instruments and effects), and Apple Loops. The Media area, therefore, is an important tool for keeping on top of these various components and is permanently located toward the right-hand side of the Arrange window. Once open, the Media area can perform a vital role in a number of media-based activities: from the drag-and-drop importing of audio files directly into your project, to the navigation of plug-in settings and the creation of sampler instruments in the EXS24.

Alongside the Media view, Logic can also use the right-hand side of the Arrangement window to display a series of list-based information. Although not as glamorous or intuitive as some of the other editors, these list views are still

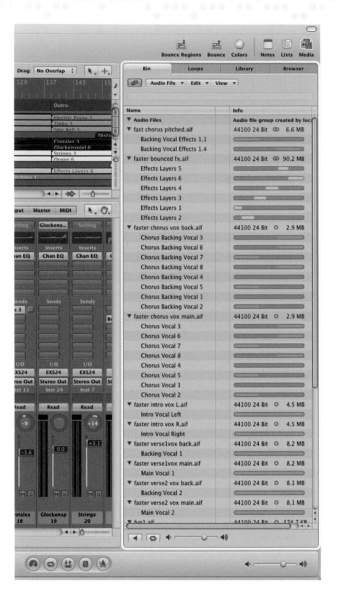

Figure 2.10
Using the Media area, you can work dynamically with the range of media files associated with a project.

vital to the success of the project and the flexibility and precision of working with Logic. In total, there are four types of list available. Event presents MIDI information in its "raw" form, allowing you to see precise parameter values and positional data. Marker displays a list of locational reference points (like verse, chorus, middle eight, and so on), which can be an effective way of navigating long projects. The Tempo list manages a project's tempo changes, particularly vital when working to picture, while Signature offers a similar control in relation to time signature.

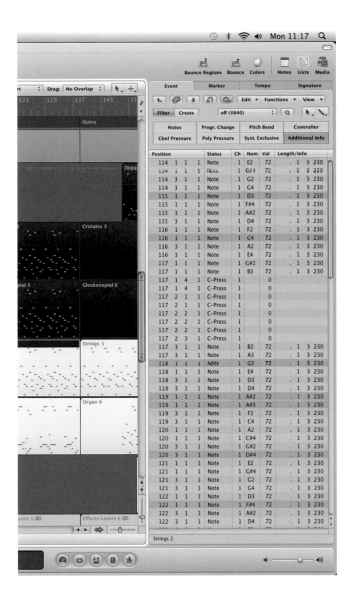

Figure 2.11 The Lists area manages the various list-based processes in Logic – from the detailed viewing and editing of MIDI information in the Event list through to the Tempo list's set of tempo changes.

2.6 Inspector Area (Keyboard Shortcut – I)

The Inspector area – found toward the left-hand side of Logic's Arrange window – provides a quick-and-easy access point to key parameters relating to the regions and tracks used within your project. The Inspector is a powerful tool for audio production work in Logic, performing many essential tasks like quantizing on MIDI regions or fades and crossfades with audio regions, as well as being an excellent way of viewing smaller parts of the audio mixer.

The Inspector is divided into three principal areas, which, looked at from top to bottom, are the region parameters, track parameters, and the arrange channel strip (see below for more information). It is possible for both the region and track parameters to be minimized, in which case you'll need to open them (using the small arrow) to access the complete parameter set.

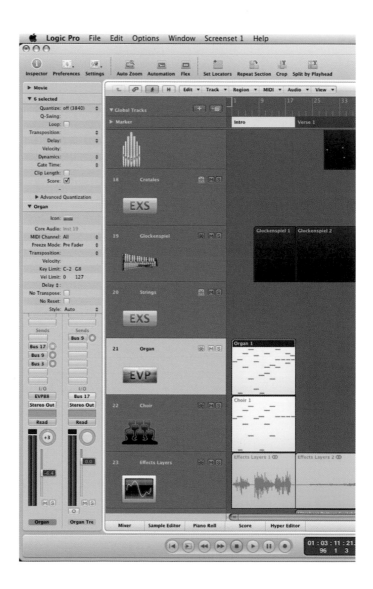

Figure 2.12 The Inspector provides a quick access point to parameters relating to audio regions and tracks, as well as a track-by-track display of the audio mixer.

Region Parameters

Region parameters govern the playback data of a particular audio or MIDI region in the arrangement. For example, a MIDI region can be quantized, transposed, delayed, and looped, all directly from the region parameters box. An audio region, on the other hand, has its own unique set of parameters, largely in relation to the use of fades and crossfades. Besides being applied on a region-by-region basis, the region parameters box is also a powerful tool for modifying playback data across several regions at the same time – maybe applying a massed quantizing setting, for example, for all the MIDI regions in your arrangement.

Track Parameters

Track parameters establish a number of parameters in relation to the track. Audio tracks tend to have slightly fewer parameters, whereas MIDI and virtual instrument tracks present slightly more control.

Arrange Channel Strip

Not a full mixer as such, but the Inspector's channel strip view at least provides some basic access to the audio mixer on a track-by-track basis. The second channel strip illustrates the "next step" in the channel's signal path. For example, this could be the main outputs, or an aux fader, if the track was being fed to reverb or delay plug-in via a bus send.

2.7 Transport Bar and Toolbar

Framing the extreme top and bottom of the Arrange window is the toolbar and transport bar, respectively. The toolbar offers quick access to important menu-based functions in Logic, from basic settings and preferences to various editing tools like region nudging, region splitting, automation, and strip silence.

The transport bar includes the all-important transport controls (play, rewind, fast forward, and so on), a display area (including SMPTE time, CPU usage, and so on), and finally, a number of Mode buttons to activate functions like the metronome and cycle playback.

To match your preferred way of working, it's worth noting that both these bars can be redesigned as you see fit by Ctrl-clicking on part of their grayed-out surface. In both bars, you can add further functions, and in the case of the toolbar, change the relative positioning and arrangement of the buttons.

Figure 2.13 Use the toolbar to access features only usually available via keyboard shortcuts or the menu.

Figure 2.14 Navigate your project and invoke various Mode buttons via Logic's transport bar.

2.8 Tools, Local Menus, and Contextual Menus

With many of Logic's principal audio and MIDI editors, alongside the main arrangement area, you'll find a series of tools dedicated to carrying out different activities or tasks. These tools can be configured for the left- and right-hand mouse button using the small drop-down menus in the top right-hand corner of the respective area. For example, the Arrange area includes tools that relate to the manipulation of regions, as well as automation data. When you're editing automation data in the Arrange area, therefore, it might be beneficial to place the main pointer tool on one mouse button, with the Automation Curve tool on another.

Pressing the Escape key at any point will also bring up the particular toolbox menu for that area, allowing you to quickly switch tools without having to travel to the extremes of the area's window.

Alongside Logic's global menus, you'll also find a set of local menus unique to each area, many of which contain powerful editing features and functions that go beyond what can be achieved by mouse alone. Although not essential, it's worth familiarizing yourself with these menu features over time.

In many cases, it is also possible to access these features via the use of contextual menus, directly activated by Ctrl-clicking in the relevant area. As you'd expect, the precise role and content of the contextual menus changes between areas, but you'll also find that the menu adapts given the precise part of the screen or objects you click on. For example, in the Arrange area, the contextual menu will differentiate between Ctrl-clicking in the main region area, regions themselves, the bar ruler, and the track list, with different functions appearing each time. Although bewildering at first, as you get more used to these features, you'll find the contextual menus a powerful and immediate means of accessing the real depths of Logic's editing prowess.

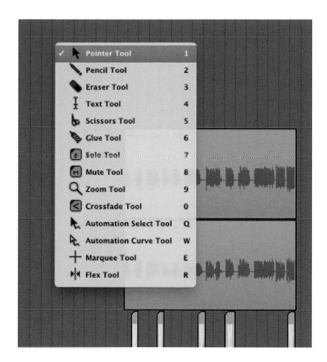

Figure 2.15 Each area has its own unique set of tools, which can be quickly accessed via the Escape key.

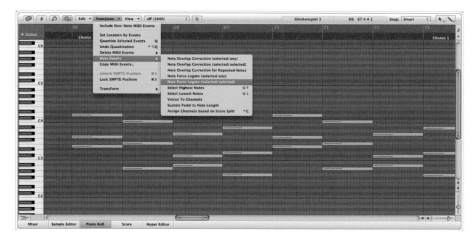

Figure 2.16 Each area will have its own local menu, with a range of functions that are unique to the activity taking place.

2.9 Adjusting How You View the Arrangement: Zooming In and Out

Varying the magnification of your project is essential both in terms of keeping an overview of the "macro" arrangement of your track and to be able to hone in on certain details, even down to sample-level accuracy. In the arrangement area,

Figure 2.17
Contextual menus – by Ctrl-clicking on the screen – access many features that are unique to the area or object you click on.

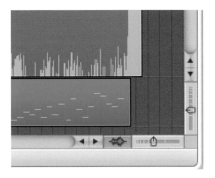

Figure 2.18
Using the various zoom functions allows you to understand both the overall structure of your arrangement and the smallest details within it.

the two zoom controls (for the horizontal and vertical axes, respectively) can be found toward the bottom right-hand corner of the window – note that you can also use Ctrl+Alt+ any of the arrow keys to achieve the same movement in and out of magnification. You can also use the zoom tool to drag-enclose a specific portion of the screen you want to view, with a double-click taking you back to your original magnification level.

As an alternative to zooming in and out, you can also use the menu option View > Auto Track Zoom (keyboard shortcut Ctrl+Z) to enable a track-dependent horizontal zoom. This allows you to better view the track that you're currently

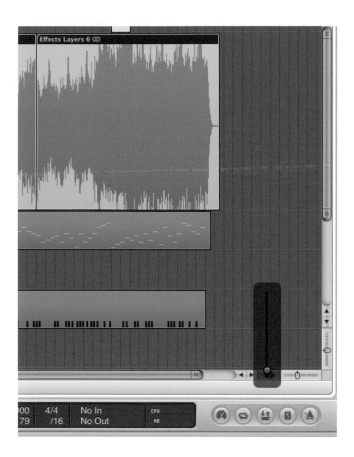

Figure 2.19
Changing the scaling of the waveform will allow you to improve your audio editing in relation to quieter parts of the signal.

working on, although, of course, this is only in relation to the horizontal rather than vertical axis.

Another type of magnification is the waveform intensity. This effectively changes the volume scaling of the waveform display, allowing you to better see quieter, more discrete parts of the waveform that you usually wouldn't be able to see. To expand the waveform, simply click and hold on the waveform icon next to the horizontal/vertical zoom controls, and use the slider to scale accordingly. Although the waveform might end up looking distorted, the meters should, in fact, confirm that it isn't.

In This Chapter

3.1 Introduction   27

3.2 USB Devices   27

3.3 FireWire Devices   32

3.4 PCI Express   34

3.5 Other Audio Equipment   35

3.6 Integration into Logic – Audio Preferences   39

3.7 Hard Drives   42

3.8 Control Surfaces   43

3.9 Distributed Audio Processing and External DSP Solutions   46

3.10 What Is MIDI?   46

3.11 Digidesign Hardware Integration with Logic   49

Knowledgebases

Latency and Monitoring   51

ReWire   53

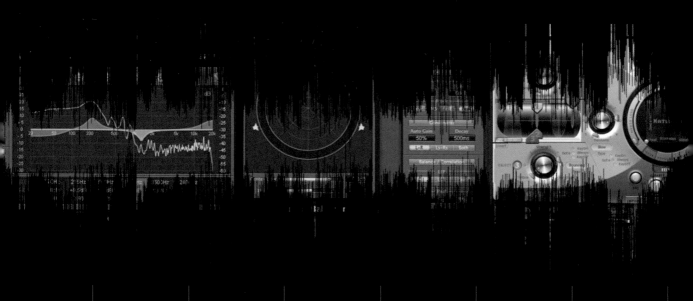

Getting Connected

3.1 Introduction

Although it's easy enough to produce music solely using your computer, it's inevitable that you'll want to add some additional equipment to enhance your workflow: whether it's a USB controller keyboard to control one of Logic's many virtual instruments, an additional hard drive for storing samples, or an audio interface so that you can record electric guitar. Understanding the various connections in and out of your computer is therefore vital to getting the best from Logic and will allow you to expand your setup from a basic home computer to a full-sized professional studio!

In this chapter, we're going to take a closer look at the key components you might want to add to your existing Logic and Mac setup – understanding the pros and cons behind their selection, how they connect to your computer, and how these devices integrate into the Logic working environment. Of course, it's important to remember that no two setups will be exactly the same, so you should select the equipment that is most appropriate to your budget and the type of music you want to produce.

3.2 USB Devices

Whilst the Mac's onboard audio facilities are respectable, they can vary considerably from model to model. Most Mac laptops come with a microphone built-in, and again whilst very respectable for speech when chatting online, it lacks the professional quality you'd expect for recording music. Whatever the onboard input, you'll soon want to simply plug things in and expand your Logic Studio. In this section, we'll look specifically at the USB devices to get you connected.

DOI: 10.1016/B978-0-240-52193-0.00003-4

USB – An Introduction

Connecting devices to the Mac is now a much easier affair due to the advent of USB. Before USB, different manufacturers tended to use varied, and often proprietary, connection protocols. USB brought this all together with version 1.0, essentially resulting in one connection protocol for a multitude of different uses.

USB 1.0 came as a slow protocol of 12 megabits per second (Mb/s), which was ample for devices such as mice, keyboards, printers, MIDI interfaces, etc. Early Macs and some devices you will come into contact with (such as some older audio interfaces) are still USB 1.0. However, USB 1.0 would slow with too much traffic via multiple connections using a USB hub.

FireWire was readily available at the same time, and being 400 Mb/s, it handled many of the heavier duties such as external hard drives and larger audio interfaces. FireWire is still the preferred choice for many audio professionals.

Fortunately, USB 2.0 came along around the year 2000 and offered a considerably faster connection speed of 480 Mb/s allowing the USB to be fast enough for hard drives and audio interfaces with higher track counts and higher data rates.

USB 3.0 was formally ratified in 2008 and will, it is expected, start to appear on computers in the coming year or so. This has a data rate said to be 10 times faster than USB 2.0.

In practice, many external devices and peripherals for computers these days are connected using the USB, including many used for music making. USB is a flexible system that enables multiple devices to be connected to one computer with the use of a USB hub. Each hub acts as a splitter/merger box and typically allows for between 4 and 10 devices to be connected at once. The USB limit is much larger than this, but is unlikely to be exceeded by most users. Let's take a look at the types of USB devices we can use with Logic Pro 9.

USB Audio Interfaces

Audio interfaces come in many forms, but typically, USB-based devices are self-powered and historically handle two inputs and two outputs. In some instances, there may be useful MIDI sockets found here too, which we'll discuss in a little detail later in the chapter. These usually inexpensive units may not always offer the highest grade sound quality, lowest latency (see Knowledgebase 1), or the largest range of sample frequency and bit-depth options, but they are excellent solutions for where basic input and output (I/O) or portability are the key. Ideal uses for this type of interface are simple stereo-pair recordings and overdubs. A large-scale live recording would require a larger interface with a more comprehensive I/O count, but to track up, a large work could be easily built upon with this type of interface using overdub techniques.

Figure 3.1 The Cakewalk UA25EX by Roland is an inexpensive audio interface with an impressive number of features for musicians on the move. *Image courtesy of Cakewalk, part of the Roland Corporation.*

There is a huge range of USB audio interfaces available and choosing the right one can be difficult. Interfaces vary in price, depending on sound quality and features. Manufacturers such as M-Audio, Edirol, Cakewalk, and others make a wide range of excellent inexpensive interfaces for this purpose, such as the Cakewalk UA25EX, shown below. Cakewalk is now a division of Roland, a long standing and respected music manufacturer, and as such, this unit typically is well built and has lots of features such as inputs for different types of microphones (including phantom power), guitars, and other connections; higher sample rates such as 96 kHz and bit depths of 24 bits; an internal limiter; optical digital inputs and outputs; and MIDI IN and OUT, a fairly impressive little device.

Apogee, a manufacturer covered in more detail later in the chapter, has recently introduced the ONE, which is a small USB interface. The ONE has a microphone built-in plus an input for an instrument or a favorite microphone. The ONE comes with the professional quality output you'd expect from Apogee.

The key features vary from device to device; hence, choosing the right device for you will be dependent on the type of music you make to some extent. If you mainly produce electronic music, it is likely that a two-in and two-out device will be ample. However, if you are in a band and want to track your next album of live performances, you'll require something with more inputs and flexibility. Many devices can handle multiple microphone inputs and also include some additional features such as software mixing and digital signal processing such as compression and effects.

USB version 1.0, unlikely to be adopted on many new devices, was a very slow connection and many of the stereo interfaces were restricted to 44.1 kHz and 16 bits. Most devices are now USB 2.0 and adopt the data rate of 480 Mb/s, which is a little faster than that of FireWire 400. Therefore, while in theory any USB 2.0 audio interface could handle a large I/O count, there would be interruptions for other devices connected to the computer (mice, keyboards, printers, etc.); thus, it is sometimes considered a little less reliable. This is due to the

Figure 3.2 Apogee has teamed up with Apple to produce some excellent audio devices. The smallest is the new ONE, which has a mono input of a built-in microphone, jack input, or external microphone. *Image courtesy of Apogee Electronics Corporation.*

data throughput originally not being as stable as FireWire is at present. As such, FireWire tends to be adopted for larger audio requirements or even PCI Express (PCIe), which are both covered later.

USB Controller Keyboards

To play in notes through Logic's instruments, you will use the Caps Lock Keyboard provided by pressing the Caps Lock key on your Mac, but sooner or later, you'll want to escalate to your own controller keyboard. Many keyboards come with MIDI sockets (see the MIDI section toward the end of this chapter) to connect to your computer, but recently, these keyboards have connected directly using USB.

The introduction of USB controller keyboards negates the need for a dedicated MIDI interface if you mainly use the internal Logic Instruments. However, if you wish to use some older MIDI equipment, you'll need to use a MIDI interface, and for more details on this visit page 46. M-Audio's Oxygen8 v2 shown below is a good example of a portable and flexible controller keyboard offering both

Figure 3.3 M-Audio's Oxygen8 v2 is a common and practical USB keyboard for musicians on the move, offering both a couple of octaves as well as transport and mappable controllers.

Figure 3.4
M-Audio Axiom
49. A four octave
keyboard with
a whole host of
controller controls
and drum pads for
music creation.

two octaves of input in addition to some assignable controller knobs. With a keyboard like this, you can not only enter in the notes but also control some of the parameters of the Logic Instrument such as the filter and envelope settings (covered later in this book).

Many controller keyboards are available across the product ranges, working up to some pretty hefty weighted numbers. The top-of-the-range controllers are often built to feel like a real piano's action under your fingers and are solid workhorses. There are many examples in between such as the M-Audio Axiom 49, which additionally includes controllers and drum pads.

Select your controller carefully depending on the music you create and the flexibility you need. Dance music production would probably benefit from a controller keyboard with assignable knobs and sliders to various synthesizer and sampler parameters, while a virtuoso keyboard singer/songwriter might prefer to ensure the feel of the keys are as close to a real piano as possible. Portability will be a key aspect and this may heavily influence your choice.

3.3 FireWire Devices

FireWire audio devices have to some extent become the standard for professionals on the move. The reason for this is due to the aforementioned speed and reliability issues of USB 1.0. As such, FireWire became traditionally used for the data hungry devices such as audio interfaces and hard drives. FireWire remains popular as its ports are rarely shared with many other devices that could interrupt the data flow.

FireWire audio interfaces remain popular, ranging from the inexpensive Mackie Onyx Satellite (shown below) with two inputs and up to six outputs, to more expensive units with a higher count of inputs or outputs. These devices are ideal for the larger projects you might wish to get involved in. Typically, these devices offer somewhere between four and eight microphone inputs and up to 10 or sometimes more analog outputs. In addition to this, there are often a multitude of digital connections, such as the ADAT lightpipe standard, which can add another eight inputs and outputs to the interface with a suitable convertor. This method of expansion is typical with many audio interfaces, and if paired with a digitally connected console or multi channel convertor unit, such as Focusrite's OctoPre MkII, most of them can be made of the units full input and output count.

Apogee is a well-respected, high-end audio converter manufacturer who is now working closely with Apple. A result of this cooperation has seen the

Figure 3.5 Mackie's Onyx series of PreAmps and audio interfaces begin with the Satellite model, which integrates a docking station for working at home, but can be removed to work on location. *Image courtesy of Loud Technologies Inc.*

introduction of three high-end audio devices. These Macintosh-only audio interfaces contain some excellent and specific benefits for Logic users. From the start, these units boast Apogee's reputed high-quality converters, which after all are a key part to the sound quality of your studio setup among many other key features.

The first in the trio is the Ensemble that boasts eight analog outputs and eight audio inputs, four of which are very low-noise digitally controlled microphone preamps, two of which contain insert points for connection of a favorite outboard processor, and the other two also allow for the connection of high-impedance instrument connections for a bass or electric guitar. The Ensemble has extra facilities for digital connectivity using ADAT lightpipe to your studio setup, which extends the I/O count to 16.

Apogee provides its own software interface between Mac OS X's Core Audio and their hardware called Maestro. This is comprehensive and offers the user the ability to configure multiple headphone mixes and outputs plus many clocking options and other intricate and useful controls. The Ensemble digitally

Figures 3.6 and 3.7 The smallest is the new "ONE" (shown on page 30). The Duet is a stereo-based interface, which like the "ONE" includes a precision digital encoder or rotary knob for assignment to controls within Logic. The larger rack-mountable device, the Ensemble, offers up to 36 inputs to the Mac. *Images courtesy of Apogee Electronics Corporation.*

controls its microphone preamplifiers and this can be managed by mouse using the software, but ingeniously, Apogee has also included a knob on all their interfaces, which they call a "precision digital encoder" for the purposes of controlling faders and knobs on screen.

The second device is the smaller and more portable Duet, which is essentially a two-input/two-output portable device. The accompanying breakout cable allows for six connections: two XLR microphone inputs, two jack instrument inputs, and two line-level outputs. The Duet also comes with Apogee's Maestro software.

Naturally, there are a plethora of FireWire audio devices on the market for use with Logic, which all should work with Apple Core Audio. However, before taking the plunge, it might be worth checking on the manufacturer's Web site that the device you are considering is compatible with Core Audio and Logic.

3.4 PCI Express

Perhaps you have an existing studio with an analog console or are replacing another multitrack system with Logic. Either way, to obtain a truly large set of inputs and outputs, a top-flight system will be required using a MacPro's PCIe Slots.

These types of systems bypass the FireWire and USB connections and instead connect to a PCIe circuit board that sits within your computer and connects directly to the computer's central processing unit (CPU). This enables much faster data rates than FireWire or USB 2.0, therefore offering larger track counts, higher sample rates, and larger word counts.

There are a few good examples of this model. One is the MOTU HD192 system, which can be expanded to accept 48 inputs and 48 outputs concurrently and works with a range of software.

Another popular PCI-based system is Digidesign's HD system using ProTools. Logic can make use of this hardware through the use of its translation software allowing Logic to access the Digidesign Audio Engine (DAE), or in other words its audio interface, directly. In addition, ESB allows Logic users to route natively processed sounds into ProTools' time division multiplexing (TDM) bus accessing the DSP hardware and plug-ins available in ProTools.

Apogee, with their Symphony interface (shown below), and RME, with their Multiface II interface, create PCIe cards to allow appropriate connections to their devices, which also offer large track counts.

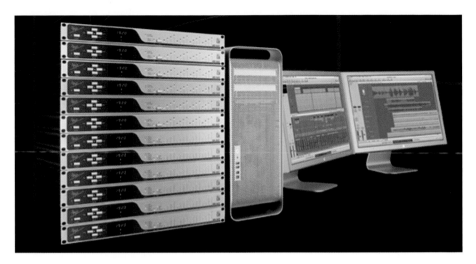

Figure 3.8
Apogee's
Symphony Series
connects to the
Mac using the
PCIe connections
and offers high
sample rates and is
highly expandable.
*Image courtesy of
Apogee Electronics
Corporation.*

3.5 Other Audio Equipment

As you expand your set up, you'll need to consider other things that connect to your audio interface. These devices are typically microphones, loudspeakers (called monitors in professional circles), headphones, and an ancillary outboard. In this section, we'll look at these in detail.

Microphones

There are many microphones on the market, and some are indeed collectors items. However, you'll need to know a few simple facts to find the right choice for you and get them connected. Essentially, there are two common types of microphone, or the way in which they work internally: the dynamic microphone and the condenser microphone.

Both condenser and dynamic microphones have their own unique benefits, and this can be used to great effect depending on what you're recording.

Condenser microphones are very sensitive, being able to capture really quiet sounds as well as handling loud ones too. Condenser microphones can also capture a wide range of frequencies from very low bass through to the very top treble. The cost of these relatively fragile microphones was for many years out of reach of most people, but recently, a wide range of condenser microphones are very affordable. Condenser microphones require some form of power source to operate. Typically this is known as Phantom Power, and it may be worth ensuring your audio interface can deliver this when making a purchasing decision. phantom power sends a 48-V DC signal down the

Figure 3.9
Neumann microphones have been an icon of vocal recording for some time and the U87 shown here is a perfect example.

microphone cable to power typically two things: the diaphragm and the internal preamplifier.

Electret condenser microphones differ slightly in that a charge is built in to the capsule containing the microphone diaphragm. As such, only power for the onboard preamplifier is required. These can, like traditional condenser microphones above, run with phantom power, but electret's main benefit is that they have a provision for battery power onboard when no phantom power is readily available. A classic example of this would be the AKG C1000S shown on the opposite page.

Dynamic microphones are traditionally less expensive and offer users the ability to record anywhere. Their characteristics include the ability to accept loud sounds well (close miked drums, guitar cabs, and the like), ability to reject spill as they're not as sensitive as their condenser counterparts, which is of great use when recording drums, for example, and their frequency range is typically also a little less too than a condenser. One other attribute is their sturdiness.

Connecting each type of microphone will be typically via the XLR connections provided on your audio interface. It is worth again pointing out that condenser microphones require phantom power (+48 V), which will need to be engaged once you've connected your microphone. Ensure your speaker levels are down as this can cause quite a large bang.

Figure 3.10 The AKG C1000 is a popular example of an electret condenser microphone which has capacity for a PP3 microphone onboard to power the onboard preamplifier.

Figure 3.11 The SM58 from Shure is possibly the most common of all dynamic microphones offering legendary solid construction and performance for vocalists live.

Monitors

Monitors are loudspeakers to you and me. In professional terms, they allow you to "monitor" the signal, hence their name. There are literally hundreds of products on the market currently and choosing a pair of monitors that suit your music and your budget will require a little investment of time.

Studio monitors can range from expensive (and often large) far-field studio monitors through to smaller and cheaper, near-field monitors.

The studio monitor market has blossomed in recent years. As a result, there are many excellent units to choose from. Many monitors represent exceptional value for the money, and care needs to be taken when choosing them.

Most monitoring systems these days are what we call "active," meaning that they come with specially developed amplifiers which work best, or tuned for that loudspeaker unit. Active speakers can represent the best performance for the money. "Passive" monitors require you to obtain an amplifier to power them.

Figure 3.12 Controlling volumes in studios where a mixing console is not necessary can be resolved using Logic's Master Output Fader or investing in a specific solution such as Mackie's Big Knob.

Some active units also come in what is known as a 2.1 configuration. Don't be put off by the numbers. These just relate to two monitors (left and right stereo) and the ".1" refers to a subwoofer, a speaker unit designed to deliver bass typically below 150Hz. It is likely that you've heard of 5.1, which relates to the surround sound format commonly used at home for surround sound. Logic can allow you to mix in 5.1 (see Chapter 10). This relates to five full range speakers (left, right, center, rear left, and rear right) and the subwoofer ".1." There are increasing numbers of speakers that are added to this, such as 7.1 and 10.1, for other surround speaker applications.

Connecting active monitors to your set up should be as simple as connecting them to your audio interface. However, it is worth noting that many active monitors do not have conventional volume controls. As such, you will need to control the monitor level from either Logic using the master fader, from your audio interface (if a level control is provided), or by some other means, such as a mixing console if you have one. Specifically designed units such as Mackie's Big Knob can offer a neat solution. The Big Knob is more than just a volume control. It offers management of headphone mixes, talkback, and routing to other devices.

Headphones

Studio headphones (also known as "cans") are another important element of the studio. They are really not very different to consumer headphones that you use already at home. Often, studios will provide what are known as closed headphones, meaning that sound from the outside is rejected as much as possible and vice versa.

Recording acoustic music without good closed headphones can lead to either the headphone mix spilling out and being heard in the vocal recording or the loud drums or guitar stack you're recording being heard over the headphone mix.

Additionally, how loud they operate can be an issue if you're a drummer and want to hear the rest of the band in the cans, but good closed designs should prevent the need for too high a level.

Figure 3.13
The DT100 headphones from Beyer Dynamic have an impressive history spanning decades. The timeless appeal is due to its closed design and replaceable parts.

A good example of a studio set of headphones is the Beyer Dynamic DT100 shown above, which has been used in studios across the world for over many years.

Outboard

From time to time, you may wish to add an analog processor such as a favorite compressor or effects unit. Doing so in an analog studio is a simple matter of patching it in, but integrating this in Logic, you may wish to consider how this might be connected.

For example, if you wish to record using the favored compressor as an insert, you may wish to see if your audio interface can accept inserts (such as with the Apogee Ensemble). Otherwise, connecting this device will be via the inputs and outputs of your interface, allowing you the opportunity to route the signal through the processor and back into the computer again. This may cause delay in the form of latency unless you're able to use onboard routing software such as Apogee's Maestro.

These inputs and outputs could also be employed as an auxiliary send and return in the case of an external effects unit. Latency should not cause too much of a problem in such instances as this may be masked dependent on the effect employed.

The expansion of your studio in this way will depend on the amount of analog inputs and outputs you have at your disposal.

3.6 Integration into Logic – Audio Preferences

Apple's operating system contains a neat feature called Core Audio. Core Audio should mean that when you launch Logic, each connected audio interface appears in the audio preferences. In fact, Core Audio and Logic allow for an interface to be connected to the system mid-session, or "plug-and-play" as it is known. Logic alerts you that a new interface has been added and asks you whether you would like to use it, as in the example below.

To check on your interfaces and settings, simply go to Logic Pro > Preferences > Audio, where the Devices tab should be present. Here, there are three subtabs called "Core Audio," "DAE," and "Direct TDM (DTDM)." Core Audio will very likely relate to the device you have just connected and is the first tab open to you. The other two tabs relate to Digidesign hardware, whose integration with Logic we'll discuss later.

Looking at the Core Audio tab, there are many parameters that can be altered to get the best from your audio interface(s), which are called Devices in Logic. The Devices list will allow you to select all the Core Audio-compatible devices available to the system at the time. Simply select the one you wish to use, and presto, you should be up and running. Core Audio makes it that simple.

Getting the best from your device will usually mean managing the latency you experience. As we discuss in the Latency and Monitoring Knowledgebase, there are workarounds regarding monitoring. However, the computer's driver and interface will also require time to "buffer" the information together before sending it out. The I/O buffer size is measured in samples and this therefore relates to the amount of time the computer and device take before you hear the sound. In an ideal world, we'd want this buffer to be at its lowest setting of 32 samples, which at 44.1 kHz would be considerably less than a millisecond and a perfectly acceptable delay. Doing this will require a very powerful computer, as more of the processing power is needed and may result in your audio dropping out or distorting. The ideal is to find a happy medium where the delay is not too long and the computer's processing power enables unhindered audio. If latency causes problems with your recordings, you can use the Recording Delay feature to move the audio back or forth so that it fits with the

Figure 3.14 When a new audio interface is added, Logic will notify you and ask you whether you want to use this device. If this does not appear, then it might be worth obtaining the latest drivers for the interface from the maker's Web site.

rest of your music. The default for this is 0 ms but can be moved by 5000 ms either way.

Bit rates in digital audio refer to how fine the measurement of the amplitude, or loudness, of the audio signal is. CD recordings are presented at 16 bits and were considered resolute enough for the domestic market. However, within music production, 16 bits is considered a little too coarse to the ear and as such 24 bits is now desirable, as it measures the amplitude scale in finer detail. Therefore, by increasing the detail and accuracy of the waveform, it is more representative of the original audio. To record using 24 bits, select the 24-bit Recording check box within this pane. Remember, though, that your audio interface will need to respond to 24 bits to benefit.

Higher bit rates than 16 bits are desirable, but come at a slight price. The data this occupies on your disk is larger than that of plain 16 bits and as such you should consider whether you have enough hard drive space before the session (covered in this chapter). Roughly, you'll need nearly twice as much hard drive space as with 16 bits.

The Independent Monitoring Level feature for "Record Enabled Channel Strips" is very useful while recording, as it gives the user control over the monitor level for that channel's input when in record mode. For instance, if you were to record a guitar amp close to you, it would be less necessary to hear a version of it from Logic in your headphones. Simply turn it down when in record mode and track away.

The Software Monitoring check box will usually be enabled as Logic presumes that you will be doing most, if not all, of your audio routing and management inside as in the example above. However, if you wish to use your studio's mixer or your audio interface's monitoring features, then this should be unchecked.

The Process Buffer Range refers to the amount of time you will allow the computer to gather and process information into the buffer before playing it. The lower the range, the faster the computer will respond and the lower the latency. However, the computer may be unable to process all the required information in the time specified and drop out altogether.

ReWire, which is covered later in this chapter, is like an internal audio and MIDI patch lead connection between two pieces of software such as Reason and Logic. ReWired Behavior refers to the type of work you are doing between Logic and the ReWire device and hence the burden on the computer. The choices are between playback and live operation. Playback assumes that the computer is being played and can use the buffer to its best effect to ensure timely playback of instruments, and this process requires less CPU load. However, live mode assumes that you are playing the instrument live and as such requires the connected device to be more timely when producing the sound; the computer needs to prioritize this and therefore more CPU load is used.

3.7 Hard Drives

Recording audio is a data-intensive exercise, and often as modern track counts increase, the internal, original hard drive of your Mac may start to strain under the pressure. As such, many Logic users have become accustomed to installing additional hard drives in their MacPro. However, for many Mac users (especially those using laptops), for reasons of space, file management, or file security, you may wish to choose to employ an external hard drive.

Choosing an external storage device will depend entirely on what you wish to do. For example, if you just want to back up your system, or your audio files, then the drive can be relatively low speed and perhaps lower cost. However, should you wish to use this addition as your dedicated audio drive, then speed and capacity could both become issues worth factoring in.

Audio files can take up some considerable hard drive space and as such a large drive is welcome to contain all your projects. Drive speed is also an issue. It is preferable for the audio files to come off the drive as fast as they went on, and as such many manufacturers, such as LaCie, provide an option for a 7200-rpm drive instead of the slower speeds such as 5400 rpm.

Although USB 2.0, rated at 480 Mb/s, is faster than FireWire 400, its structure and ability to hub mean that it can share its valuable bandwidth with other devices such as mice, keyboards, cameras, and MP3 players. As a result USB is, at times, considered to be less reliable for data efficiency, throughput, and stability. Thus, FW400 appears to remain the de facto standard despite it not being the fastest presently.

However, FireWire devices currently come in two flavors: 400 Mb/s and 800 Mb/s. Most devices still utilize the common FireWire 400 (FW400) format due to its history, popularity, and reliability. More hard drives and devices are now utilizing the newer FW800 format, which can offer very fast data rates and is becoming more popular for audio duties.

Hard drive manufacturers such as LaCie make a series of drives suitable for the Logic user, from portable, self-powering drives such as their Starck range through to the industry-standard d2 Quadras. The d2 series has been the Logic professional's choice for the main part and now supports a multitude of interface formats, usually USB 2.0, FW400, FW800, and the new 3-Gb/s eSATA connection all on the same unit.

SATA and Internal Hard Drives

Serial Advanced Technology Attachment (SATA) is the connection protocol between a computer and its internal hard drives. This is a very fast method of connection working at speeds of either 1.5 Gigabits per second (Gbit/s) or newer 3 Gbit/s. When working in a MacPro, it is possible to connect up to three

Figure 3.15 The LaCie d2 for many has become one of the industry-standard hard drives of choice, offering both reliability and flexibility (including a rack mount available from LaCie). *Image courtesy of LaCie.*

additional hard drives to that of your system disk. For audio and multimedia, this is an excellent way to work as different drives can be used for differing projects or for backup purposes. Separating out the system drive containing Mac OS X and Logic from the audio allows the drives to be dedicated to specific tasks. However, the simple overriding benefit is the speed of SATA drives for working with high data rates such as multitrack audio.

3.8 Control Surfaces

The studio was traditionally a tactile environment where physical faders, tape machines, cables, and instruments were the norm. As the digital revolution has taken its hold, so much of that studio has been pared down to the computer screen. As such, the notion of grabbing the fader for any channel to make an on-the-fly adjustment becomes a little bit more involved, usually requiring several mouse clicks perhaps with some keyboard presses too. Computing power and software development has meant that the amount of features controlled from one mouse and keyboard has grown dramatically. A common solution to this problem is to bring back the tactile aspect of the studio in the form of an external control surface.

Control Surfaces come in many forms from the one fader units such as the Presonus FaderPort to the fully fledged Mackie Control Universal Pro. What is common is that most come in the form of a bank of faders with some control for other aspects of the channel strip such as pan and auxiliary sends. As you navigate around the controls, Logic should also follow suit and vice versa, giving you tactile control over your mix.

Control surfaces such as Mackie's original Control Universal are connected using the normal MIDI cables, although the newer version (shown below) and other systems, such as the FaderPort, adopt the USB protocol. Other controllers are combined with audio interfaces as one package such as Roland's SI 24, thus replicating the function of the mixing console in its entirety. These and many other control surface devices are connected to the computer using USB, FireWire, and on rare occasions, Ethernet.

Figure 3.16
Mackie's Control Universal Pro has been the choice of many Logic users due to its impressive integration and overall flexibility.
Image Courtesy of Loud Technologies Inc.

Figure 3.17
PreSonus' FaderPort is a simple and intuitive control surface allowing adjustment of one fader at one time, plus the usual transport controls.
Image courtesy of PreSonus.

Setting up a control surface is easy, provided they integrate with Logic natively using an existing or accompanying driver. To install a new device, go to Logic Pro > Preferences > Control Surfaces > Setup …. You'll face the Setup pane that has three menus at the top: Edit, New, and View. Select New > Install from the menus at the top, which will open another window listing all the Logic-compatible control surfaces. You are encouraged to "Add," or you can scan for attached devices.

Once you have added your controller, close this window and return to the controller setup page. Here, it is possible to select the controller from the diagram on the right-hand side and then edit any information about the device in the inspector-like space on the left-hand side. Within this space, you're able to manage many aspects and preferences, such as what MIDI ports it is connected to and what feature the fader controls within Logic.

It is possible to add more than one model of control surface. As such, Logic makes it possible to edit what the controls actually operate in Logic within the setup page. In the example of more than one controller, duplicated buttons such as play and stop could be reassigned on one unit, while the faders could naturally lead on, say from Channel 8 on one device to Channel 9 on the next control surface. To get two control surfaces working in tandem, ensure in the Control Surface Setup window that the controller pictures form a horizontal row. However, should you wish for the controllers to act independently, then place the icons in a vertical column.

The way in which the Control Surface reacts and interfaces with Logic can be comprehensively edited using the Controller Assignments pane (Logic Pro > Preferences > Control Surfaces > Controller Assignments … or simply press cmd + K). There are two views here: Easy View and Expert View. In Easy View, each command can be seen one at a time and it is simple to reassign a control to another duty. To do this, simply press on the part of Logic you wish to control. Next, press Learn Mode to move a control or touch a button on the controller surface to map it accordingly.

To edit these modes, choose Expert View; the area to the left of the pane describes the zone of the surface, in this example the Mackie Control Universal. The control surface itself is split into zones such as VPots, Tracks (faders), Global Views, etc. The next table refers to the modes in which each zone can operate, and by selecting one of the modes, the Control/Parameter list will change to show you what each control does.

The Controller Assignments editor is incredibly powerful as any feature of the control surface can be remapped to another to allow you perfect control over a particular session or mix. In the example below, the selection list can be seen on the right-hand side offering a large number of choices for remapping an EQ control.

3.9 Distributed Audio Processing and External DSP Solutions

As with any project, there will come a time when the onboard processing power of the computer will start to strain under the pressure of plug-ins within your mix. As with Digidesign's HD ProTools system, there is an external solution to the problem for Logic users too. Distributed audio processing allows for an additional DSP to be made to the computer either by Ethernet, such as Logic's Nodes, or by an external DSP solution connected via FireWire or PCI.

Using a system such as Solid State Logic's (SSL's) Duende or Universal Audio's UAD system, power can be unleashed from your computer to make for a faster and more powerful workflow. The audio processing is taken care of outside the host computer using proprietary processing to each manufacturer. As such, plug-ins can be purchased to add to the systems, which might be more reliant on your DSP than you would desire for your computer.

Setting up an external processing device such as SSL's Duende is easy and is often a matter of simply connecting the FireWire and installing the accompanying driver. With OS X rebooted, Logic should now pick up that it has Duende installed. Logic may rescan its audio units and an extra line within the audio units menu when selecting plug-ins called Solid State Logic. Within here should be the features that are available on the external unit such as the SSL channel strip.

For more installation guidelines, please refer to the manufacturer's instructions and Web site.

Using Multiple Macs

Distributed audio processing can be achieved by connecting multiple Mac computers via their Ethernet connections. These slave computers become what are called "nodes" in Logic speak and will allow for your processing tasks to be distributed across a unique network. Setting up nodes and multiple Macs is covered in Chapter 11. This is very flexible as it allows you to harness the power of your older Mac or desktop workhorse, but still allows you to take your laptop out on the road.

3.10 What Is MIDI?

MIDI was originally developed as a communication protocol between hardware synthesizers, samplers, drum machines, and of course, sequencers like the original MIDI-only version of Logic. The protocol is made up of a number of different types of messages – including note on, note off, velocity, modulation, and so on – all of which can be sent to 16 available MIDI channels. In theory, the channel system was originally used to denote different devices along a "daisy

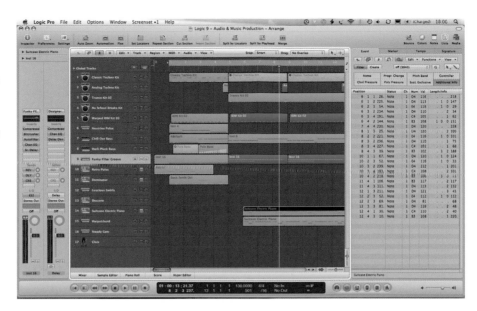

Figure 3.18 The MIDI information can be clearly seen within either the separate Events List window (Windows > Events List or Apple + 0) or the Events Pane in the Arrange window by pressing 'E'. The Events List shows the corresponding MIDI information contained within the selected MIDI region.

chain" of MIDI leads, with each device sifting its appropriate collection of message from the collective stream of MIDI data passing along.

Nowadays, of course, hardware MIDI devices are much less common (although you may well use a MIDI controller keyboard plugged into the MIDI port of the Audio Interface), but the MIDI protocol is still alive as an effective means of "internally" communicating with virtual instruments in Logic's audio engine. Open up the Event list, for example, and you still get to see the list of familiar MIDI events, all with their associated channel and parameter numbers. Also, the Instrument Parameters in the inspector allow you to specify a MIDI channel for the instrument in question (although most default to ALL), and of course, instruments like the EVB3 use different MIDI channels to differentiate between the two different keyboard manuals and the bass pedals contained within the instrument.

So, even though you may never connect a MIDI lead again in your life, MIDI is definitely here to stay as an integral part of sequencing with Logic. It is really important to appreciate MIDI as your communication language when controlling instruments within Logic.

However, if you're itching to use an older classic MIDI synth, you'll need to connect this to your set up. As we've established, many audio interfaces come with MIDI sockets. However, if you do not have these or if you wish to connect a multitude of synths together, you may need to consider a dedicated MIDI interface, which connects to your Mac's USB connections.

Figures 3.19 and 3.20 Mark of the Unicorn's (MOTU's) Fastlane is a solid and inexpensive USB MIDI interface, while their MIDI Time Piece AV is packed with additional features. *Image courtesy of Mark of the Unicorn.*

Many manufacturers fulfill this function. Mark of the Unicorn (MOTU) is one such manufacturer, which offers a wide range starting with the Fastlane (a two-in and two-out MIDI interface) all the way through to the professional MIDI Time Piece AV (eight-in and eight-out, with many additional features).

To the rear of these interfaces lurk a variety of five pin sockets known as the MIDI sockets. These usually come labeled in three ways: IN, OUT, and THRU. As you might imagine, MIDI IN receives instructions from another device – for example, the sound module will receive an instruction from the MIDI OUT of Logic's MIDI interface. If the sound module sends information back to Logic, such as MIDI controller data, then the MIDI OUT will need to be connected to the MIDI IN of the interface. These MIDI connections apply to every MIDI-compatible device within your studio.

Additionally, MIDI THRU is included in the MIDI protocol to allow for something that is known as daisy chaining. In the very early days of MIDI (pre-Logic), it was unlikely that your sequencer would have more than one MIDI output, two at the very most. As such, if you had more than one MIDI device, it was important that you could connect it to the sequencer with ease. Enter the MIDI THRU socket, which replicated the MIDI IN data and passed it through the device allowing you to connect it straight to the MIDI IN of the next device unaltered. On rare occasions, some MIDI devices would merge both the MIDI OUT and MIDI THRU of a device so that a continuous loop back to the sequencer could be made. In this day and age, multiples of MIDI INs and OUTs are possible using modern MIDI interfaces as discussed earlier; therefore, the MIDI THRU socket is less used.

3.11 Digidesign Hardware Integration with Logic

Digidesign, the manufacturer of ProTools, produces a number of audio interfaces that roughly fall into two categories. The first of these is their LE range, which is a host-based system, meaning that all the digital signal processing is handled within the computer's processor. These plug-ins are known as a Real Time Audio Suite (RTAS). The second category is intended for professional applications. Digidesign pioneered the TDM system that makes use of the DSP power held on their external HD cards, leaving the host CPU to get on with the work at hand.

Both of these methods can be used with Logic as an audio interface using Digidesign's Digi CoreAudio Manager. The Digi CoreAudio Manager effectively bridges between the hardware and OS X's Core Audio Services, allowing you to control the hardware as though you were in ProTools Hardware Setup page. When using the Digi CoreAudio Manager with Logic, it is worth noting that Logic and the LE interface may not immediately talk to each other. Opening up Digi CoreAudio Manager before launching Logic should ensure communication.

Digidesign's HD range requires you to connect the external audio interfaces to their HD, PCI-based cards placed within your desktop computer. It is quite common to find both Logic and ProTools coexisting on the same computer, as they have traditionally been used for differing applications at separate times. They can coexist with Logic directly accessing the DAE and therefore the Digidesign hardware. The processing marriage between Logic Pro and the Digidesign DSPs is handled by the formerly named Emagic System Bridge, now simply "ESB TDM" to access processing from the ProTools domain. (TDM is Digidesign-speak for the type of processing that utilizes the processing power within their DSP chips inside the HD card.)

Logic does not normally host such 'external' technology. However, it is possible to access the Digidesign hardware as though it was an extended part of your Logic family. Accessing that DSP processing power on the Digidesign HD cards would be useful. This is achieved by using two mixers: the TDM Mixer employs Digidesign hardware to process TDM plug-ins whilst the DTDM Mixer gears up the Mac processor to manage the Logic host AU plug-ins. For more information on how this integrates, it is advised that you refer to the Logic 9 Pro TDM Guide from the Logic manual (link below).

To set up Digidesign hardware with Logic, go to Logic Pro > Preferences > Audio > Devices > DAE. Simply select the Enabled checkbox and you'll be asked to reboot to ensure that the hardware is all present and correct. This will allow you to connect to the Digidesign HD hardware that is attached. You should also tell Logic how many HD cards and interfaces are attached by clicking on

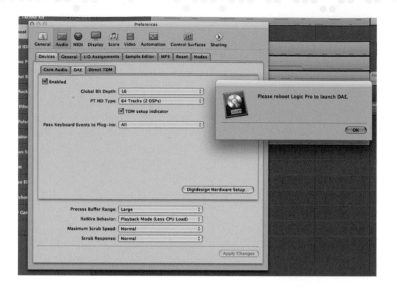

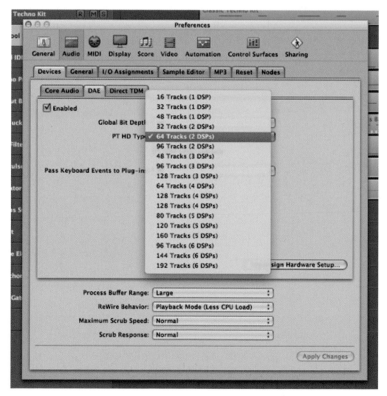

Figures 3.21 and 3.22 Setting up Digidesign hardware can be achieved within the audio preferences pane.

the ProTools HD Type list (PT HD Type). Within this list, there are details of the ProTools track count determined by the hardware and the amount of DSP chips that are included on the HD card.

A good place to start when working with Digidesign hardware is to open up the TDM Configuration template from the Template Chooser. This offers you a mixer and tracks, which will work with a Digidesign setup and allow you to get started. By employing this template, you can soon see how Logic interfaces and works with the Digidesign hardware.

For more information on Digidesign HD integration with Logic, see the dedicated Logic Pro 9 TDM Guide, which can be found at http://www.documentation. apple.com/en/logicpro/tdmguide/#chapter=preface%26section=0.

Knowledgebase 1 ▼

Latency and Monitoring

When recording was in the analog domain, the signals fed from the mixing console through to an open-reel multitrack tape recorder and back would only be delayed by an acceptable millisecond or so. To all intents and purposes, this was perceived as pretty much instantaneous, certainly within acceptable margins. As this delay increases, performers can be put off their performances and they might appear to be out of time.

This delay in modern digital audio is called latency and is a common side effect of working with digital audio workstations such as Logic. Latency is caused by the delay experienced as your computer, audio and MIDI interfaces process the information. A common example of this is the delay between pressing the MIDI keyboard and the sound generated from an audio instrument in Logic.

Overcoming latency can often be tricky as it depends on the speed of your audio interface and computer. Monitoring is one major issue to overcome, and this can be achieved by employing an external mixer to handle your headphone mix. Many inexpensive USB interfaces combat latency by employing a "mix" knob on the unit, such as Digidesign's MBox. This allows you to blend the input signal with the output of the computer, which allows you to hear your input signal prior to any incurred latency. Some Audio Interfaces come with their own software-based audio mixer, which performs this monitoring duty (such as MOTU's CueMix, shown on the next page), as well as often integrating onboard digital signal processing.

(Continued)

If one of these solutions is used, then it is desirable to switch off the Software Monitoring checkbox in Preferences > Audio. Another solution is to employ Logic's Low Latency Mode.

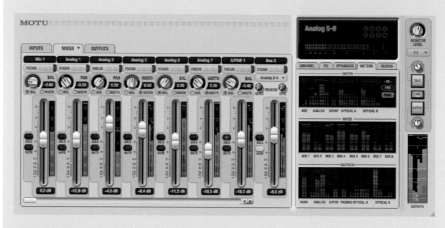

Figure 3.23 MOTU's CueMix is an excellent example of the additional mixing features that can be provided specifically for the audio interfaces. *Image courtesy of Mark of the Union.*

Latency can be problematic when performing using a virtual instrument or monitoring a recording. Logic has created a neat solution called Low Latency Mode which intelligently bypasses active plug-ins on the channel you're working on, thus freeing up processing power. This allows for latency to be managed to the preset limit of delay you have chosen in the preferences. To change this limit, go to Logic Pro > Preferences > Audio > General. Turning on Low Latency Mode is easy as there is an icon directly on the transport bar. In most cases, this will alter the sound of the channel you're working on, but this is only temporary for the time this feature is engaged.

One of the best current solutions to overcome latency with Logic is to use Apogee's Symphony hardware connected via PCIe to the Mac. This boasts some of the lowest latency available at the time of writing.

Knowledgebase 2 ▼

ReWire

ReWire is a system developed by Propellerhead, the makers of Reason and ReCycle. ReWire cleverly allows internal connections to occur between audio software packages within one computer. It is therefore possible to have multiple MIDI connections internally between applications and receive the corresponding audio signals back in return, again internally. This is an incredibly powerful and flexible solution when getting the best from two applications. Setting up ReWire in Logic is very easy. You must ensure Logic is opened before the ReWired application is started.

With Logic opened, launch the ReWire application, thus ensuring that Logic knows that this is to be the ReWire application and will therefore make the appropriate connections for you. To use your ReWire instruments, first begin to create an External MIDI track by pressing cmd+Alt+N on the Arrange area. Next, call up the Library Tab from the Media Icon on the top right-hand side of Logic. Within the Library should be your opened ReWire application – in this instance, let's say Reason. Clicking on the Reason folder in Logic's library should reveal all the open Reason instruments. Simply click on one to map this to the selected MIDI track.

As you select new instruments within your ReWire application, both the MIDI and audio connections should be automatically made to Logic. The audio connections will appear as Auxiliary tracks on your Logic Mixer and can be processed using Logic's plug-ins and blended into the mix as though they were another virtual instrument.

In This Chapter

4.1 Introduction   55

4.2 Assets and Projects   56

4.3 Working with Tracks   58

4.4 An Introduction to the Audio Mixer   60

4.5 Using the Transport and Timeline   63

4.6 Your First Recording   65

4.7 Overdubs and Punching-In and -Out   67

4.8 Creating Further Tracks and Track Sorting   70

4.9 The Audio Bin and Importing   72

4.10 Working with Apple Loops   73

4.11 Improving What the Artist Hears – Headphone Mixes   77

4.12 Monitoring through Effects   80

Knowledgebases

What are Assets?   57

More About the Metronome   66

Importing REX2 Files   75

Plug-in Focus

Amp Designer   82

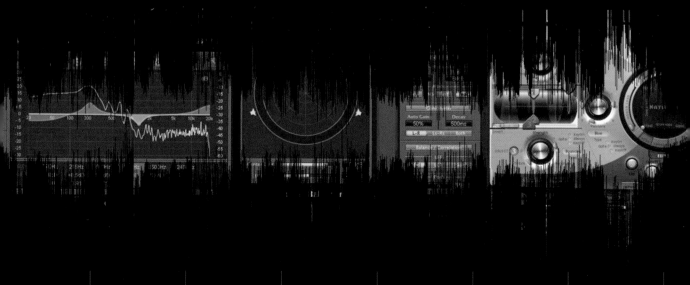

Starting a Project

4.1 Introduction

Laying the successful foundations at the start of your project is vital to the long-term success and seamless workflow of realizing audio and music production work in Logic. Of course, not all projects start the same way, but we're going to begin by looking at the process of recording audio into Logic – in effect, using it as a modern day replacement to the multitrack tape machine. This is also an excellent way of covering some important essentials of working in Logic, like creating tracks, using the transport controls, the bar ruler, and basic monitoring. We'll also start to explore the various aspects of data management involved in working with Logic – more specifically, how and where you should store your project data, and what a Logic project is actually comprised of.

If you're short of inspiration or just needing something to kick-start your creative process, we'll also take a look at importing and working with Apple Loops. Apple Loops – contained within the multitude of Jam Packs installed with Logic Studio – are an instant resource of loops, instrument, and vocal passages, all copyright-free and ready to be used in your compositions.

Besides the basics of track laying, we'll also review various ways with which you can make a recording session run even more smoothly – either providing a dedicated headphone mix for musicians, for example, or using plug-in effects to better understand how your instrument might eventually appear in the mix. All these factors will create a more comfortable recording experience for you and your musicians, allowing you the musicianship, engineering, and performance qualities to become more focused.

Of course, this is only the starting point to your creative work in Logic, as we'll also be looking at how to edit this newly recorded material in Chapter 5, and then about integrating the range of virtual instruments and MIDI sources in Chapter 6.

4.2 Assets and Projects

The basic currency of Logic is a project, which will contain all the relevant assets for the song you're creating. In respect to Logic, the term "assets" corresponds to the collection of separate files and data that a song can be composed of. For example, in a song file, which contains all the important arrangement information, you could easily assemble any amount of audio files, sample data (for the EXS24 sampler), and movie files, all of which will reside in the main project folder. By organizing your data in this way, both you and Logic can keep better track of your creative process, making backing up, or moving to a different Logic 9 setup, for example, considerably easier.

The starting point of your creative process is to create a new project (File > New). Logic will then present you with a series of possible starting points, including some interesting options for composing or producing music in Logic, either based on genre or production objectives, like surround mixing, for example, or mastering. If any of these templates meet your particular needs, then it might be appropriate to select them, otherwise select the Empty Project as a suitable initialized starting point for production. Note that you can also automatically select this option by holding down Alt as you select New from the File menu. Assuming that you've selected an Empty Project as your starting point, Logic will prompt you to create an initial track to work with. To get the ball rolling, select Audio under the track Type option and press the Create button to insert a single track. We'll take a deeper look at creating tracks in Section 4.3.

Figure 4.1
Creating a new project (File > New) will present a number of starting templates. Select Empty Project if you want a simple initialized starting point.

Save As...

Before you go any further, it's worth saving your project, if only to allot a particular location for your project folder. If Logic hasn't already opened a Save As dialog, you'll need to do so (File > Save As …). Besides specifying the name of your project, you can also instruct Logic to automatically save the relevant assets to your project folder, under the Include Assets option. It's also worth checking the advanced options, as this will allow you to differentiate between whether Logic simply collates your respective audio files or added extras like EXS sample data, instrument files, Impulse response, movies, and so on. Ultimately, if you're unsure, there's no harm in selecting all the options, but this will use a correspondingly larger amount of drive space.

Knowledgebase 1 ▽

What are Assets?

Assets are Logic's collective term for the various data files that can be associated within a project. Of course, the project file – or song – is only the beginning of the data you'll amass over time, with the full set of assets including audio files, movie files, EXS24 sampler instruments, ultrabeat samples, and space designer's impulse responses. Ultimately, if you don't have some organized strategy for looking after these files, you'll soon find that a project becomes difficult to manage, especially when it comes to either archiving a project or moving it to another system.

By keeping the files in a single folder, Logic offers at least one way of organizing this data, although if you've got an alternative system, you can always decide to manage this yourself. Theoretically, this can prevent certain unnecessary file duplication, which can easily become an issue with large EXS24-based projects. To change the current project's Asset settings at any time, select File > Project Settings > Assets, where you can, accordingly, switch in and out the different parts of the asset management. By default, Logic concentrates on audio files, including automatic sample rate conversion of files imported that don't match the current project settings.

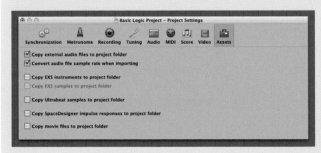

Figure 4.2
Managing the assets related to your project will allow you to create better project archives as well as move between different Logic setups in a more effective way.

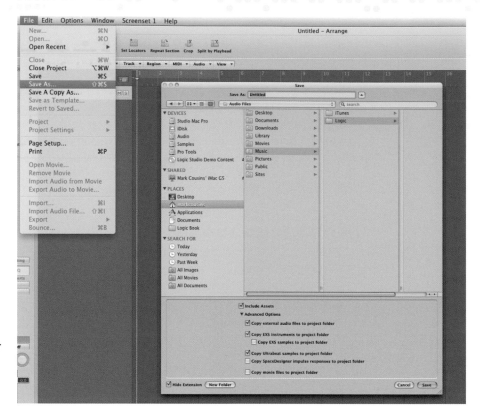

Figure 4.3 The Save As dialog allows you to create a root folder for your project, as well as to check what assets will be saved with it.

Although this initial project management can, initially, appear time consuming (especially if you just want to get on with making music!), it's worth pointing out that good data management is a vital part of music making on a computer. Spending time to ensure that you know where your data is being stored will reap plenty of dividends in the long run, making it far easier to recall projects several months down the line, or indeed, work more effectively between different computers.

4.3 Working with Tracks

Assuming you've selected an Empty Template, the next task is to create a series of tracks to record your audio and MIDI information into Logic. For example, if you're recording a band using a number of microphones, you'll need to create an accompanying collection of audio tracks representing each instrument or microphone you're using. If you intend to create a track largely using virtual instruments, with a few vocal overdubs, you'll want to create 16 or so instrument tracks combined with one or two audio tracks. Of course, you can always

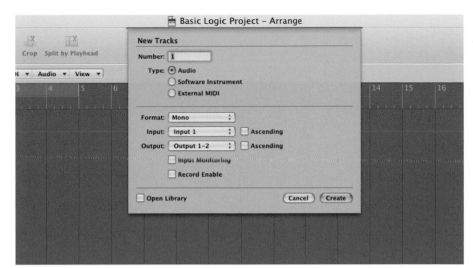

Figure 4.4 Create a number of empty tracks to correspond with the instrumentation you want to record. The New Tracks feature includes several ways of speeding up this process.

add, remove, or reorder tracks from your project at any point in the production process, so don't feel obliged to preplan your complete track list at this point.

To create a track, click on the small + sign on the top of the track list, or, if no tracks are present, Logic will ask you what tracks you initially want to create. The dialog box that accompanies track creation allows you to specify both the exact type of track that is being created, as well as several options for creating multiple tracks, output and input assignments, and so on.

Audio

An audio track will allow you to record an audio signal being fed from a corresponding input on your audio interface. This is the primary focus in this chapter.

Software Instrument

Use a software instrument track if you want to record music using any of Logic's integrated virtual instruments (EXS24, EVP88, ES2, and so on), or indeed, any third-party Audio Unit instruments. We'll be taking a closer look at the process of working with virtual instruments in Chapter 6.

External MIDI

External MIDI tracks correspond with MIDI hardware like synthesizers and samplers that you have externally connected to Logic (usually through a USB interface). Again, we're going to spend more time looking at this functionality in Section 6.5.

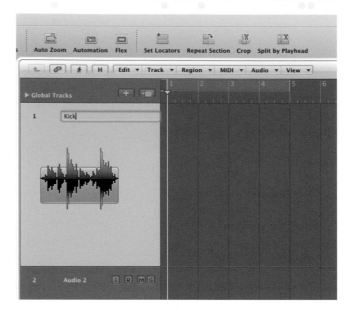

Figure 4.5
Naming tracks will allow you to keep a better handle on your Logic arrangement, as well as being used to automatically name the audio file and region that you create.

As you're creating your first set of audio tracks, you'll notice a couple of important options to speed up your workflow. For example, if you intend to record from the first 16 inputs of your audio interface, you might wish to increase the number (of track) parameter to 16, and click on the Ascending Input option – that way, each subsequent track will take a different input number. On the whole, most users tend to monitor their session from a single output, so you'll probably want to leave the output setting out of its Ascending option. Whether its format is mono or stereo will depend on the type of signal input being fed to your audio interface – a pair of drum kit overheads, for example, will be stereo, whereas a vocal (recorded with a single mix) would be mono.

With your basic tracks created, you'll want to also create some initial track names. Not only will this make the navigation of Logic's Arrange window and Mixer much easier, but you'll have better chance of keeping track of the various project audio files with names like "kick drum" rather than "audio 01." To name a track, simply double-click on the existing track name as part of the track list. Once named, any subsequent recordings on that track will adopt the name accordingly.

4.4 An Introduction to the Audio Mixer

Although we don't need to go too deeply into mixing at this point, it is worth introducing the concept of the audio mixer, mainly as an alterative way of looking at the tracks you've created for your project. While the track lanes of the

Figure 4.6 The Mixer window (accessible from the Mixer tab or by pressing X) displays an alternative view of your newly created tracks.

Arrange area work horizontally, the Mixer area shows us our current project in the form of a traditional vertical mixer, which each track represented as a different channel strip. You can open the Mixer at any point using either the Mixer tab at the bottom of Logic's Arrange window or from the keyboard shortcut X.

Figure 4.7

Common to both the Mixer and Arrange area's track list are a series of function switches, each of which govern how the particular track or channel is working at any given point in time.

Input Monitor (I)

Input monitoring is a simple one-click means of directly monitoring, or listening to, the current selected input for that given track or channel. At first, this might be a good initial way of establishing what's coming into your inputs, and, using the meters on the channel strip, setting an initial gain on your microphone and line inputs. Note that in most cases, this will be done directly from the interface itself (or the interface's control software, if you're using something like the digitally controlled Apogee Ensemble) rather than adjusting the fader's position. In this case, the fader is simply setting the monitor level, rather than any form of recording level as it might do on a conventional console.

Should you wish to change a track's particular input, this can be done from the Mixer, clicking on the small input and output (I/O) box above the fader. The list presented should tally with the inputs available on your audio interface. For example, a typical USB audio interface will present two available inputs, whereas a FireWire interface could offer 10 or more possible inputs. Clicking on the small circle button on the bottom of the fader will also change track between mono and stereo operation, with a corresponding change in the input selection – input 1, for example, would become input 1–2.

Figure 4.8 Use the I/O box on the channel fader to change a track's input status, with the number of inputs corresponding to the type of interface you're using with Logic.

Figure 4.9

Record Arm (R)

Record arming a track or channel indicates to Logic that you're about to make a recording. Again, you'll be able to see your input levels accordingly, and hear the signal routed through to your main outputs. Once you press record on the transport, Logic will actively record the input as a new audio file (more on this in a moment!).

Figure 4.10

Mute (M)

As you'd expect, mute silences the track or channel in question. What can be confusing, though, is how this differs between the Arrange area and the Mixer. As the controls are independent (unlike the record arm or Input Monitor options), you won't find changes on one reflected on the other. Instead, use mute to silence the track or channel. For example, you might run several track lanes for the same instrument (like different takes of a vocal) where mute would silence one of those. Muting the channel, however, would silence any signal being directed to that strip.

Figure 4.11

Solo (S)

Solo will play the chosen track or channel in isolation, and, like the mute function, works independently so that you can either solo an individual track lane or a number of track lanes being fed to the same channel strip.

Working between the Mixer and the Arrange area is an important part of the day-to-day usage of Logic. For example, note how a selection in either window is mirrored in the other – selecting the vocal track on the Arrange area, for example, will also bring it up on the Mixer.

4.5 Using the Transport and Timeline

As with any multitrack recorder, the transport bar – found toward the bottom of the Arrange window – is what provides the master control for recording, playing back, and looping in Logic. For example, having created your tracks and record-armed them, pressing Record on the transport will tell Logic to engage its playback (of any existing material) and record each track as an individual audio file. The rest of the controls (play, pause, rewind, fast-forward, and so on) need little introduction, although it's worth noting how the transport controls are also influenced by settings in the Arrange area's bar ruler.

On the right-hand side of the transport, you'll find a number of different Mode and Function buttons – like a low-latency monitoring mode, solo, and of course, the Metronome. We'll be covering these in more detail as we move through the recording functions in Logic.

Bar Ruler

The bar ruler – along the top of the Arrange area – provides the positional reference, or timeline, for our project, with the Playhead indicating the current song position. As you'd expect, there are a number of additional functions of the bar ruler, allowing you to specify parts of the recording to be dropped in, for example, or areas to be looped around. Crucial to the functionality are the left- and right-hand locators, displayed numerically on the transport, or the lighter gray shading in the bar ruler.

You can adjust the left and right locators in a number of ways, although probably the quickest and most logical solution is to drag from left to right (with the mouse button held down) in the top half of the timeline. As you drag your selection, you'll automatically place Logic into cycle mode – as illustrated by the both green shading in the bar ruler, and the fact that the cycle mode indicator (found in the bottom right-hand side on the transport) is active. In cycle mode, you can repeat over a section of the music, and if you're recording, lay down a number of different passes or takes of the same musical line to be edited later on.

To disable cycle mode, you can either click on the green region in the bar ruler or alternatively deactivate cycle mode on the transport. Even when it's deactivated, Logic will retain the left- and right-hand locator settings allowing you to return to the cycle at any point. Furthermore, you could also define a number of different cycles using the Marker feature (like Verse, Chorus, and Middle Eight), which we'll cover in more detail in Chapter 10.

Figure 4.12 The transport bar also includes a number of important Mode and Function buttons – including the Metronome and Autopunch – that are vital to the process of recording in Logic.

Figure 4.13 Here are two left and right markers defining the length of the cycle. Use the cycle feature either to loop round a portion of the song or to perform a series of different takes.

Tip ▽

As well as using the on-screen controls, you can also use Apple Remote Control to operate the transport of Logic. Using your infrared control, you can remotely start and stop the transport, move backward and forward bar by bar, and toggle through the various track lanes. To record, simply press the Menu button – a great solution for anybody recording themselves away from their computer!

4.6 Your First Recording

Assuming you've created and record-enabled the appropriate amount of audio tracks, you should be ready to make your first recording. Recording will begin in one of two places – either at the current position of the playhead or, if you've set up cycle playback, at the point the cycle begins. To provide you with a small amount of preroll, Logic will actually start playback a bar earlier than the current record point, allowing you to get some indication of the track's speed, as well as existing elements of the performance that might lead up to it.

When you've finished your first pass, take the track out of record arm mode. If you've got multiple tracks record-enabled, this can be done by Alt-clicking on any one of the active record-enabled tracks. Playing the project back will now allow you to audition the takes, spotting any deficiencies or areas you might want to improve.

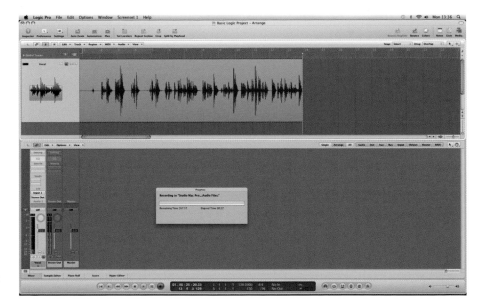

Figure 4.14
Recording the first pass into Logic.

Knowledgebase 2 ▽

More About the Metronome

In certain situations, you might find it beneficial to adapt some of the qualities of the Metronome – stopping the Metronome being sent over MIDI, for example, or changing the preset Klopfgeist sound to something of your own liking. The Metronome settings can be either accessed through the project's settings (File > Project Setting > Metronome), or directly by Ctrl-clicking on the Metronome icon on the transport bar. The settings allow you to switch the Metronome in and out of MIDI playback, as well as being able to disable the software click and adjust some basic parameters like tonality and volume.

Delving deeper in to the audio mixer environment (which we'll cover in more detail in Chapter 11, but for now access through Window > Environment) you can also change the default Klopfgeist instrument assigned to the Metronome. Klopfgeist is a simple instrument, specifically designed to produce distinctive click sounds that can be audible over a mix, yet not so loud that they would bleed out excessively over headphones. If you've got the Mixer environment open, you should see the Klopfgeist instrument assigned to the Click object. By changing this to the EXS24 sampler, for example, you could replace the standard Klopfgeist click with a rimshot sample from one of the factory kits.

Figure 4.15
Adapting the Metronome settings can be achieved through the menu option File > Project Setting > Metronome.

Difference between Mix Levels and Monitoring Levels While Recording

One aspect that you might have noticed is that Logic effectively stores two fader positions for each channel: the level for recording and the level for playback. For example, as you're recording the part, you might find it beneficial to increase the level of track for monitoring purposes – that way you can hear the instrument being recorded more clearly. In playback, however, you might want to reduce the level so as to hear the part better in the context of the mix as a whole. Move back to recording again, and the original record monitoring levels will be restored.

4.7 Overdubs and Punching-In and -Out

Although we'd all like to think we could record perfect "one-take wonders" every time we enter the studio, there's always an inevitable amount of remedial recording that needs to take place – either with fluffed notes or whole sections that need to be replaced. Logic, of course, provides a number of solutions for this, all of which can be further refined when it comes to editing your project (as we'll see in Chapter 5). Unlike tape, it's worth remembering that any punch-ins or overdubs are "nondestructive" – in other words, the original takes remain untouched on the hard disk, with Logic simply changing or augmenting the files that it decides to playback for any given track.

Quickly Punching-In Material

One solution to repairing your recording could be simply to record over a section of the track in question. To do this, all you'll need to do is place the track or tracks you want to replace into record-ready mode, locate the playhead at the punch-in point, and press record. Logic will then give you the usual bar preroll and then initiate recording, effectively "replacing" any previous recording you have made at that point (although remember, this is nondestructive). When you press stop, you'll effectively punch-out, leaving subsequent audio recordings untouched.

Recording in Cycle Mode

If you've got a specific section of the song you want to work on, it might be appropriate to record in cycle mode. In cycle mode, Logic will continually loop the portion of the song, recording each pass that you make. When you've stopped the transport, the last take will be used for playback, although all the various takes will have been recorded.

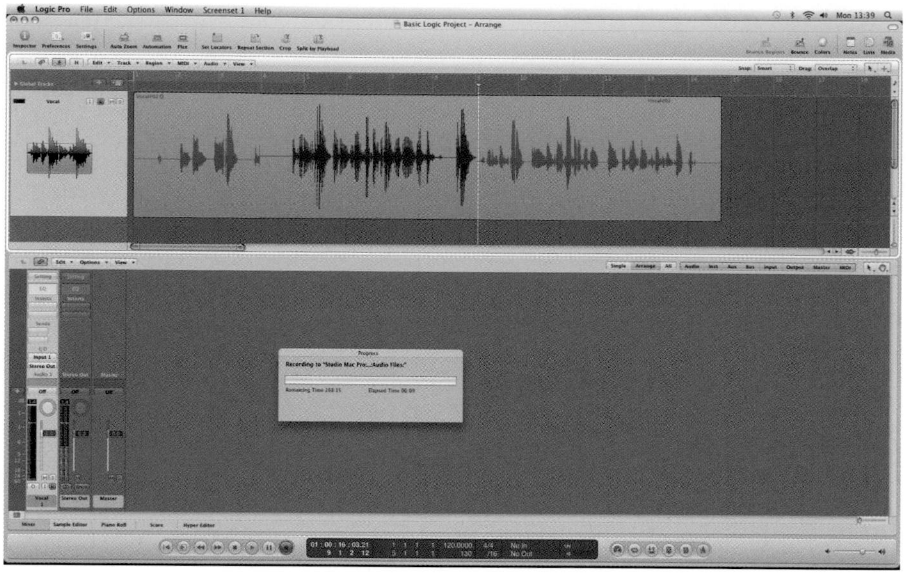

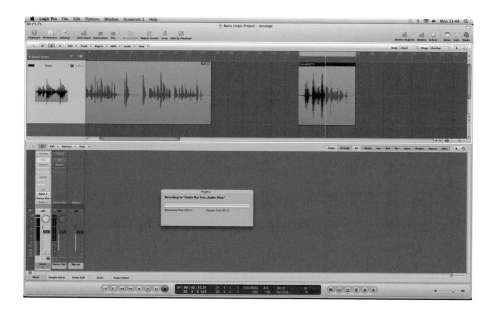

Figure 4.16 If you want to replace any section of the recording, simply locate the transport at the point where you want to drop in, and press record. Logic will provide a bar of preroll and then begin recording your new material.

Figure 4.17 Using the cycle function is a useful way of loop-recording a number of bars until the part is performed correctly.

Autopunch

If you're clear about a specific part of the track you want to replace, consider using the Autopunch feature. Autopunch can be activated through the transport bar, with an additional set of locators (shaded red) viewable in the middle of the bar ruler. You can change the position and length of the Autopunch

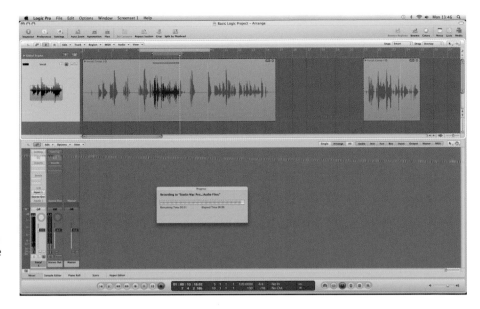

Figure 4.18
If you have a specific part of the take you want to replace, consider using the Autopunch feature for an accurate, automatic drop-in and drop-out.

Figure 4.19
If you record over an existing region, Logic will create what is known as a Take Folder. A Take Folder will allow you to reposition and refine edit points later on in the production process.

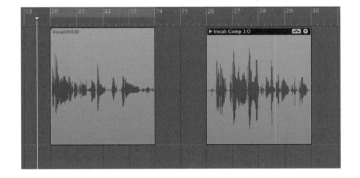

locators in the same way as the cycle locators, ideally setting the locator points to precisely match the portion of the track you want to replace. With the track record-enabled, and record engaged on the transport, Logic will playback the song and only record for the precise duration of the Autopunch locators.

Introducing Take Folders

Whenever you overdub onto an existing recording on an audio track, or use the Autopunch feature, you create what Logic refers to as a take folder. If you look closely at a take folder, it looks slightly different to a conventional audio recording in Logic, with two triangles in the top left- and right-hand corners of the region. Technically, a take folder stores all your various takes in an organized way, with features both to switch between takes and to make a quick edit between the different takes.

To understand how we can carry out further work with the take folders, take a look at Section 5.4.

4.8 Creating Further Tracks and Track Sorting

Although we should have created the majority of tracks required for our session at the start of the project, it is highly likely that you'll need to add further tracks into your session. Let's take a look at some of the motivating factors and the quick steps to create and organize these tracks.

Creating Duplicate Tracks

Use the menu function Track > New with Same Channel Strip/Instrument (or Alt + Apple + S) to create a parallel track with the same instrument or channel assignment. This is a useful tool if you want to create two alternate "takes" of the track – so, rather than packing different takes in a Take Folder, you end up with two discrete takes assigned to the same channel. Remember that only one audio recording can take priority at any given point in time, so you'll need to be clear about what track you're listening to by muting the unwanted takes accordingly.

Figure 4.20
Creating a duplicate track is useful way of negating the Take Folder system, allowing you to clearly visualize each take as it's being recorded.

Creating Tracks with the Next Instrument

Creating tracks with the next instrument or channel strip (Track > New with Next Channel Strip/Instrument) is a great way of augmenting your existing project, adding further track lanes for each new instrument you might want to add into the mix. Using the keyboard shortcut, Alt + Apple + X, allows you to do this quickly and efficiently without interrupting your creative workflow.

Creating Tracks with Duplicate Settings

If you've already set up some particular monitoring EQ and compression effects, for example, you might want to create new tracks with duplicate plug-in settings (Track > New with Duplicate Setting). The newly created track will have its own unique channel assignment, but a copy of any of the plug-in settings as well as any fader or pan positions you've established on the original track.

Sorting and Organizing Tracks

As the project grows, it might become beneficial to organize your track list into an order that better corresponds to the arrangement of instrumentation, as opposed to the order you recorded it in. For example, maybe you want to sort all the vocal tracks together with the lead vocal first, or separate the keyboard from the guitars. Arranging the track order is as simple as clicking on a track lane and shuffling it up or down the track list accordingly. Should the track list become increasingly confusing, you might also want to return it to something closer to its original arrangement. In that case, go to Track > Sort Tracks by > Audio Channel, for example, and have the tracks automatically reordered by their channel assignment.

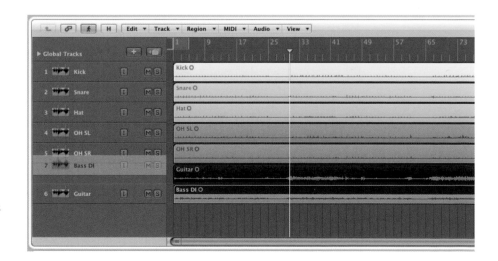

Figure 4.21
Try sorting tracks to make their organization clearer.

4.9 The Audio Bin and Importing

Although the Arrange area provides an important positional overview of the audio files we recorded as part of our project, it's not the only way we can manage and organize the various files we've created. The Audio Bin presents the complete list of the audio files included in your session, as well as offering a means of importing new audio files (possibly from another project, or a drum loop from an audio CD). To open the Audio Bin select View > Media, or open the Audio Bin using the keyboard shortcut B.

Scrolling down the Bin you should be able to see all the files that are contained within your project, even the ones that might not be currently active in your Arrange area. Alongside the file name, you can also get some basic information on the audio file, including its sample rate, bit depth, and file size. Another important piece of information is the file's location, which is displayed in the top line whenever a particular file is selected. If, as part of your asset management, you specified for all files (recorded or imported) to be stored in your project folder, then all these locations should be the same. If the project's data management isn't quite so organized, this might be an opportunity to find out which files are located outside your main audio files folder.

Importing Other Audio Files and CDs

For viewing existing audio files, you can also use the Audio Bin to import new audio material into your Logic project. Logic supports a number of different audio file formats – including WAV, AIFF, SDII, CAF, MP3, AAC, SMF, and

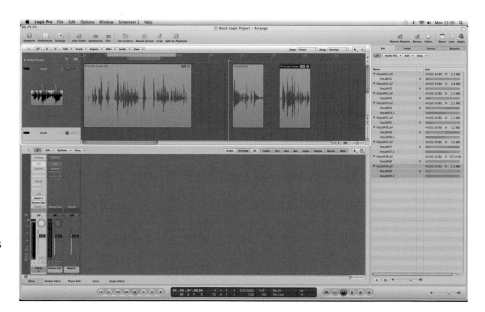

Figure 4.22 The Audio Bin provides an effective overview of all the audio files used within a project.

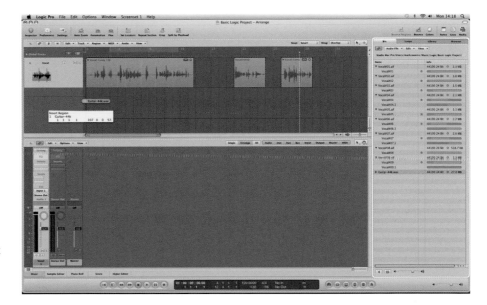

Figure 4.23 Use the Audio Bin to import additional audio files into your project. These can then be dragged into your arrangement with a new track created accordingly.

REX2 – all of which can happily reside together in the same project. To import a file into your project, go to Audio File > Add Audio File and browse your drives accordingly. Again, if you've got your asset setting established correctly, the files will be copied into your project's audio files folder to keep all the media content in a centralized, project-specific location.

To finish the importing process, you will, of course, need to drag the audio file into your project's arrangement. The precise method you choose to do this will have an effect on how the resultant file is played back. If you drag the file onto an existing audio track, the new audio file will be played back with the settings on that current channel strip. However, if you drag the new region to a position in the Arrange area without a track assignment (this would be at the bottom of the track list), Logic will create a new track specific to the audio file in question.

4.10 Working with Apple Loops

Apple Loops provides a quick-and-easy source of inspiration for music and audio production, and can be an effective way of kick-starting a project. Included with the full install of Logic Studio are a number of Jam Packs providing everything from urban drum loops to Foley effects. The important point to note about the Apple Loops format is that the tempo and/or the key of the loop will automatically conform to your project's current settings, allowing you to quickly audition a loop's relative merits without having to time-stretch the file to fit. Equally, if the project's tempo changes at a later point (even midway through the song!), the Apple Loop will follow these changes accordingly.

You can access the Apple Loop browser under the Loops tab of the media area. The Apple Loops browser is an extremely effective way of exploring the content installed with Logic Studio, allowing you to specify various search tags to sift through the huge amount of possible sound content. The browser itself can be organized into the different viewing styles – a Column view, Music view, and Sound Effects view. Arguably the most intuitive of these is the Music view (indicated by the small musical notes), which organizes the various loops by genre (Jazz and Country, for example), instrument (Vibes, Strings, and so on), and sound-type "categorization" (Melodic, Distorted, Clean, and so on). By clicking on combinations of these tabs you should be able to arrive at a loop, or collection of loops, that best suits your needs. Pressing the Reset tab (top left) will remove any existing selections, allowing you to perform a new search.

The alternative browsing methods allow you to explore the Apple Loops sound content in different ways that might better suit your methods of production. For example, the Sound Effects view (found under the FX tab) uses a variation on the music view, with a replacement set of categories that better relate to the sound to picture work. The Column view, however, sifts through the sound content in a way similar to the Finder in its column mode.

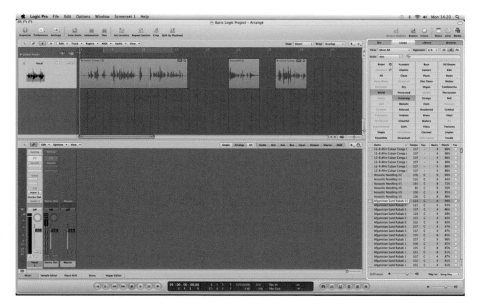

Figure 4.24 The Apple Loop browser allows you to quickly locate different loops based on their musical qualities.

Knowledgebase 3 ▼

Importing REX2 Files

The REX file format was originally developed by Propellerhead as a cross-platform solution for transporting time-sliced audio files created by their popular ReCycle application. A number of third-party loop libraries make use of this file format, so it's useful to know how Logic handles these files.

One option is to load the REX file directly into the Arrange area itself, with the slices left as a series of different regions locked to your particular project tempo. To import the loop, go to the Browser and drag the REX file into its first position in the Arrange area. Logic will then ask you to confirm the REX file import, possibly suggesting a series of crossfades to best match the original loop tempo with the tempo of your session. If you confirm, this will create a folder in the arrangement containing all the accompanying REX slices.

As an alternative to importing the slices, you can also choose to render the REX file as a complete file or an Apple Loop by selecting the appropriate option from the pull-down menu. Arguably, a rendered audio file is easier to manage than a collection of slices, although it won't be able to follow changes in the project's tempo. As an Apple Loop, though, the REX file will retain its elasticity, allowing it to match the project's tempo at any point in the creative process.

Figure 4.25 REX files can be directly imported into your Logic arrangement, with a number of different options to conform the loops to your project's tempo setting.

To hear an Apple Loop simply click on its file name, with Logic playing the file back at the same tempo and key as the song you're working with. You can also use the small volume slider and the bottom of the loops browser to set the volume for auditioning the loop. It's also worth checking the percentage match, as this will have some effect on the relative quality of the loop in the final arrangement, with loops closer to the project's tempo requiring less time compression or expansion to get them positioned correctly.

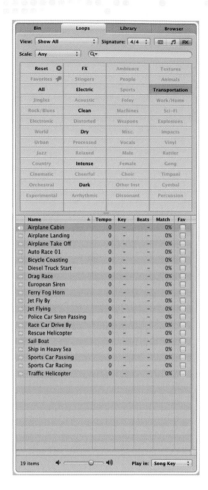

Figure 4.26 The Sound Effects view provides an alternative way of navigating and viewing your Apple Loops sound content.

Audio Files versus MIDI Files and Instruments

The Apple Loop files format actually contains two variations – a pure audio file that offers a recording of the loop in question and a MIDI version (also known as an Apple Loop +) that combines the raw note data with an accompanying software instrument and effects. You can see what type your Apple Loop is by looking at the small icon to left-hand side of the file. A blue waveform icon indicates that the Apple Loop is only available as an audio file, whereas a green music note indicates the Loop is available as raw musical data (in other words, a MIDI version) and as an audio file.

In the long run, a MIDI-based version will provide much greater flexibility than an audio-based Apple Loop, allowing you to precisely edit the musical data, instrument settings, and effects. However, these versions can, at times, make slightly larger demands on your central processing unit usage.

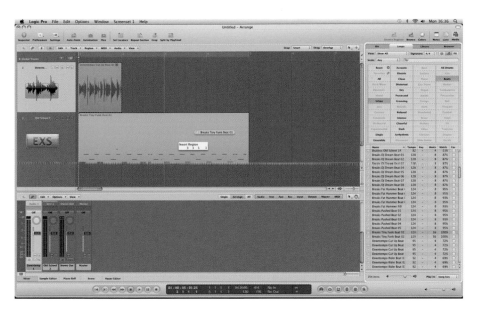

Figure 4.27 Here are the two Apple Loop file formats and how they differ when imported into your session. A standard Apple Loop (top) is audio-only, but the enhanced version (bottom) contains the specific MIDI and instrument data used to create the loop.

Importing Apple Loops into Your Session

To use either of the two Apple Loop formats, all you need to do is simply drag the file into your arrangement. One point to be clear on, though, is the type of track you import it to and how this has an effect on the end form of Apple Loop imported. For example, if you drag the Apple Loop onto an empty track lane with no track assignment, Logic will automatically create the appropriate type of track (Audio or Instrument) based on the particular Apple Loop in question. By dragging the Loop to a specific type of track, though, the Apple Loop will be conformed to the track type – so, for example, MIDI-based Loop would be placed as its audio version on a corresponding audio track. The only exception, of course, is that audio-only loops can't be converted into MIDI regions.

Once you get more proficient in Logic, you might also want to consider creating your own Apple Loops files, which allow you to access the same "elastic time" properties with your own audio material. For more information on this, see Section 5.13 in Chapter 5.

4.11 Improving What the Artist Hears – Headphone Mixes

Although we've looked the process of basic track laying in Logic from the engineer's perspective, it's worth considering what exactly the artist hears and performs to as the material is being recorded – in respect to either the individual headphone mix or the addition of plug-in effects – like compression, distortion, or reverb – that might be added during the process of recording. However, it should be mentioned that these subsequent techniques will push your Logic skills much harder than the first part of this chapter, particularly with respect to the use of the Mixer.

First off, let's consider the two types of mixes heard in a typical tracking session in a professional studio.

Monitor Mix

The monitor mix is simply the mix that is heard in the control room and is usually based on a reasonable and considered balance of the material that has been recorded. In some cases, the producer or engineer might want to hone in on a specific part (usually whatever is being recorded at that point) with the rest of the instrumentation significantly quieter. Ultimately, a good monitor mix is the only way you can ensure you're making an effective recording.

Cue Mix

The cue mix (also known as a talent mix or headphone mix) is the balance of material that relates specifically to the performer and what they hear through their headphones.

Of course, a pragmatic solution to recording sometimes involves the monitor mix and the cue mix effectively being the same thing, especially if you're recording yourself! If you're using a USB audio interface, it might also be the case that the interface features some direct provision to control the mix between the input signals and the DAW's mix so as to combat any latency issues (for more information on Latency see Chapter 3), again resulting in a combination of both the monitor and cue mixes. However, if your audio hardware supports it and you need a more professional amount of monitoring control, it's far better to create a separate monitor mix and cue mix directly from Logic.

Creating a unique headphone mix in Logic involves a more detailed use and exploration of the Mixer page, so as to effectively create two mixes at the same time. On each of the channel faders in your project, you should find a series of two empty boxes under the Send section. As we'll see later on the mixing chapter, these bus sends, as they are called, have several different applications, but for now, we're going to concentrate on their role in the creation of a cue mix. To create a bus send for a particular channel strip, click in the small grey Sends box and insert a send to bus 1. Note that if you have a large number of tracks, you can also select multiple channels, insatiate the bus on one of the channels, and have them all duplicate the same action.

The next important thing to do is to set the pre/postassignment of the bus sends. Although this sounds complicated, all a post/preassignment means is

Figure 4.28 Use the Sends box to create a bus sends for each individual channel in your mix. The bus sends are the controls that will be used to create your headphone mix.

whether the bus send is patched before or after the channel fader. Thinking through logically, it makes most sense to create the cue mix pre fade – so that any movement of the channel faders won't immediately change the relative mix for the headphones. In other words, if you want to turn down the vocal in your monitor mix, the vocalist can still hear himself or herself loudly and clearly. The pre/postassignment can be easily switched by clicking on the bus send and changing the mode from post (the default setting) to pre. If this is done correctly, the coloring of the send should change from blue to green.

In creating the bus send, Logic will have also created a corresponding aux master object, which will act as the master fader for you cue mix. To make things clearer, you could double-click on the channel's strip name toward the bottom on the Mixer and change this to something like *headphones* or *cue mix*. To build up the headphone mix, solo off this channel and then blend the appropriate instrument using the bus sends on the individual channels. As a good tip, maybe start by placing the loudest possible instrument in at around 0 dB (you can do this by Alt-clicking the corresponding bus send pot), and then positioning the rest of the instrumentation around this.

To send the headphone mix to the specific physical interface output, change the output status of the aux master accordingly.

Figure 4.29
Switching to pre fade will allow the mix to be completely independent of Logic's channel faders. The color of the bus sends indicates the pre/post status, with green being pre fade.

4.12 Monitoring through Effects

In some cases, performing a take can sometimes be tricky without an idea of how it will sound in the final mix. Although it is often the intention of the engineer to record "dry" sounds devoid of artifacts like compression or reverb (which will instead be added in a controlled way in the mix), the same cannot be said of what the artists themselves prefer to hear. Far from being fussy, effects like reverb can actively help a vocalist pitch themselves, equally the drive and distortion of a guitar sound is such an integral part of how a guitar is "played."

Effects plug-ins can be instantiated using the small gray Inserts boxes in the Mixer's various channel strips. Clicking on box will bring up a list of Logic's own internal plug-ins (organized into groups – like delay, distortion, filter, and so on), as well as any third-party Audio Unit plug-ins that you have installed on your system (organized by manufacturer). You can also activate EQ at any point by double-clicking on the corresponding EQ box just above the inserts. An important point to note is the order of the plug-ins – reflected in the progression from top to bottom on the inserts – which will dictate the order in which an instrument is processed. If you're a guitarist, for example, you might want to place certain effects before and/or after your virtual amplifier (in this case, Amp Designer). Alternatively, a vocal track tends to have EQ and compression before reverb, so that the reverb isn't equalized and compressed.

Figure 4.30
Recording through an effect plug-in will have no direct impact on what is being recorded, only what you and the musician hear in their mix.

We'll be taking a closer look at the art of setting up plug-in effects on inserts, and how their order affects the processing taking place, in more detail in Chapter 8.

The point to remember with plug-ins instantiated in this way is that they have no direct impact on the actual signal that is recorded to disk. Although this might first seem a little frustrating, it means that, if you change your mind later on, you can always return back to its initialized state – peel back the layers of distortion, on a guitar, for example, and you'll still have the completely clean, DI'd version. In effect, these plug-ins are used exclusively for monitoring purposes – they're keeping the musician happy with what they hear in monitor mix, and the engineer content with what is being printed to disk.

Accounting for Processing Latency

Although most of Logic's own plug-in are optimized for low-latency operation, it is possible that creating long chains of plug-ins, as well as using certain third-party plug-ins, might start to add unacceptable amounts of latency for real-time recording and monitoring. If playing starts to become unresponsive, you might want to consider removing certain plug-ins, although these can always be reinstated once it comes to the mix (when latency isn't an issue).

As alternative to manually switching plug-ins in and out, try using Logic's own low-latency mode, which is available directly from the transport bar, or under the General Tab of the Audio Preferences (Logic Pro > Preferences > Audio), where you can also modify the accepted latency limit from the default of 5 ms.

Figure 4.31
Using the low-latency mode will ensure that any DSP-hungry effects plug-ins won't create unacceptable amounts of monitoring delay.

Using the low-latency mode is an excellent way of being able to quickly move between tracking and mixing tasks, without having to worry about which plug-ins are causing latency problems.

Plug-in Focus ▼

Amp Designer

Amp Designer is a virtual presentation of a guitar amplifier, including the amplifier head, speaker cabinet, and virtual microphone used to "record" the amplifier's sound. As such, Amp Designer is an essential plug-in if you're recording with an electric guitar plugged directly into the instrument-level input on your audio interface (also known as the DI input). The plug-in transforms the otherwise sterile tone of a DI'd electric guitar into something far more playable – whether it's a heavily rectified metal guitar, or a clean Fender Twin tone. Although you "play through" the plug-in, the guitar is recorded in its raw state, effectively allowing you to modify the tone of the virtual amp at any point in the production process.

Although it's possible to browse Amp Designer's various presets, you can often have just as much fun swapping-out the various components (using the drop-down menu at the bottom of the interface) to create your own customized sound. An Amp Designer "Model," as such, is a combination of amplifier head, cabinet, and the most appropriate microphone. Moving between the different models, therefore, is a good way of auditioning the principle types of tone available to you.

Of course, a more detailed approach allows you to mix and match the various components of Amp Designer – maybe matching a particularly amp head with a cabinet of your choice. Note that each Amp will distort in a slightly different way – moving between the soft bluesy crunch of a Small Tweed Amp, for example, through to the rich and heavy saturation of the high octane head. The different cabinets "color" the distortion in a variety of ways, using different phase cancellations and high-frequency roll-off between the different designs.

Changing the choice and position of the microphone works in much the same way as in the real world – with a condenser microphone offering the widest and most neutral tone, while the ribbon and dynamic mics sound slightly more colored (although this is no bad thing!). Changing the position of the microphone changes the tone, with a bias toward a bassier tone as you move toward the edge of the virtual speaker.

As you'd expect, Amp Designer makes a great pairing with Pedalboard, which we'll cover in more detail in Chapter 8.

Figure 4.32 By mixing and matching the various components in Amp Designer you can create your own customized guitar sound.

In This Chapter

5.1   Introduction   85

5.2   The Big Picture – Forming Your
      Basic Arrangement   86

5.3   Using the Inspector   87

5.4   Working with Loops, Copies,
      and Aliases   88

5.5   Precise Region Editing   90

5.6   Snapping – Understanding
      Your Editing Grid   90

5.7   Resizing and Cutting
      Regions   93

5.8   Fades and Crossfades   97

5.9   Speed Fades   101

5.10 Quick Swipe Comping   101

5.11 Expanding the Take
      Folder and Creating Basic
      Comps   102

5.12 Time Slipping a Quick Swipe
      Comp   104

5.13 Flattening or Merging
      a Comp   105

5.14 Flex Time Editing   108

5.15 Flex Time View and Flex Time
      Modes   108

5.16 Transient Markers and Flex
      Markers   111

5.17 Quantizing Audio with
      Flex Time   115

5.18 Separating Regions Based on
      Transient Markers   117

5.19 Freezing a Flex Time Track to
      Save CPU   118

5.20 Editing Multiple Tracks Using
      Edit Groups   119

5.21 Drum Replacement   121

5.22 Sample Editor Principles   123

5.23 Sample Editor Applications   126

5.24 Inserting and Deleting Sections
      of Your Song   130

5.25 Working with Tempo
      and Loops   131

5.26 Managing Your Audio Files and
      Regions   137

5.27 Using Folders and Hiding
      Tracks   141

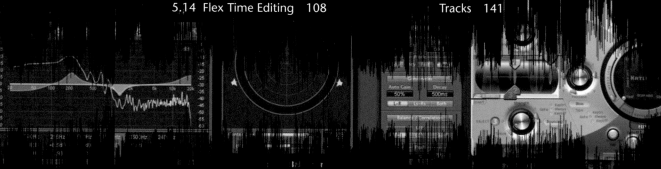

Audio Regions and Editing

5.1 Introduction

Editing and arranging audio regions has become an integral part of the recording process – whether it's simply tidying up a few misplaced notes, or creating the complex "glitch and stutter" edits that are common to many modern production styles. Therefore, in keeping with this increased importance of audio editing, Logic includes a wealth of different editing features – from Quick Swipe Comping to Flex Time editing – all of which unite to offer an almost unprecedented amount of control (and creative input!) in how your eventual production sounds to the listener.

In this chapter, therefore, we're going to take a detailed look at all the editing features that Logic has to offer. Starting off from the basics, we'll first explore how to use simple tools within the Arrange area to start forming the structure of our project – bringing parts in at different locations in the song, for example, or looping two to four bars of music to form a repetitive feature in the track. Exploring deeper, we'll look at the various tools that can be used to improve an original performance – from the ability to "comp" a performance from a number of different takes through to ways of tightening a performance and making it lock more tightly to a predetermined groove.

Of course, there's also plenty of creative potential to be explored using audio in Logic, discovering the intricacies of features like the Sample Editor, for example, or manipulating audio regions far beyond their original form. As you'll soon see, Logic offers a vast array of creative possibility in audio editing, and effective use of its editing features forms the cornerstone of contemporary music making.

DOI: 10.1016/B978-0-240-52193-0.00005-8

5.2 The Big Picture – Forming Your Basic Arrangement

In Chapter 4, we explored the basic principles of getting audio regions into your Logic project – as audio recordings, prerecorded audio files, Apple Loops, or REX2 files. However, having acquired these assets, we now need to start thinking about how we can organize and edit this raw material together to form our finished song. As you'd expect, the precise way you execute this will vary given the material you're presented with, so you'll need to be conversant with the various approaches and solutions for handling and editing audio regions in Logic.

Broadly speaking, your goals will be focused on two levels of interaction. First, you'll need to pay attention to the overarching structure and arrangement of the project, using the various editing tools to bring instruments in and out of the song. Second, you'll need to investigate the detailed qualities within each performance – like the choice of different takes at various points of the song, for example, through to the timing of each note that is played.

The first simple technique, therefore, is the ability to start moving regions within your arrangement – making copies of regions, looping regions, or muting them. You can move a region simply by selecting it and moving it to a different position in the Arrange area, either at a new point in the timeline or even onto a completely different track. Holding down Alt as you perform the drag allows you to create a copy of the region – maybe copying over a chorus

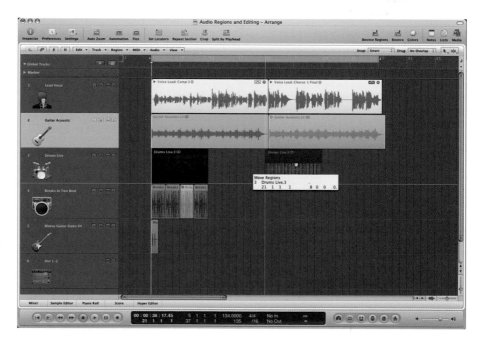

Figure 5.1
Forming the basic arrangement – either by dragging regions to new locations and copying them (by holding down the Alt key), or muting them.

Figure 5.2 The edit tools play an important part in the editing and arrangement process.

vocal line, for example, or a repeated guitar motif, so that it appears at several different points in the track.

As you'd expect, the Tool menu (which can be quickly accessed by pressing Escape) plays a vital part in the editing process. The application of most tools should be apparent, although at this stage, you'll probably want to principally work with the Pointer Tool (to move and copy regions), the Eraser Tool (for deleting regions), or the Mute Tool (to temporarily disable a region from playing back). We'll be looking at the Scissors, Crossfade, and Marquee Tools later on in the chapter, although it's worth noting their presence for now.

5.3 Using the Inspector

In conjunction with the graphical changes made in the Arrange area, it's also worth noting another important feature in Logic's approach to audio editing – the Inspector. As we first saw in Chapter 4, you can find the Inspector toward the left-hand side of the Arrange area, assuming that it has been toggled into view by selecting View > Inspector, or by using the keyboard shortcut I.

The Inspector's Region Parameters display (seen at the top of the Inspector) shows the current text-based information that accompanies any selected region. The exact information displayed in the Inspector's Region Parameters display will vary given the type of region selected, although given that we're principally working with audio regions at this stage, you should see options that relate to the Delay, Gain, Looping, and Fade settings.

Figure 5.3 The Inspector's Region Parameters display highlights the current settings for the selected regions.

The purpose of the Inspector's Region Parameters box in editing is two fold. First, the Region Parameters box provides an extremely accurate way of interacting with a region's current parameter settings – like increasing a region's gain by +1 dB, for example, or applying a fade that lasts just a few milliseconds. Second, and arguably just as importantly, the Region Parameters box also proves itself useful as a way of applying the same edit across multiple regions – like fading-in all the regions at the start of a song, for example, or time-adjusting multiple percussion parts.

Rather than covering all the applications of the Inspector here, we'll be dealing with various uses throughout the rest of this chapter.

5.4 Working with Loops, Copies, and Aliases

Many compositions feature a large number of repeated regions – like a two-bar drum loop, for example, or a repeated bassline – that occur throughout the duration of a song. In Logic, there're a number of different ways you can deal with repeated musical phrases – from the ability to create multiple "hard" copies to the option of creating instantaneous region loops that last the full duration of the track.

Making Multiple Copies

As an extension of the existing region duplication feature (using the Alt key, in other words), consider using the menu option Region > Repeat Regions (cmd + R).

This allows you to create multiple copies of the selected region/s along the duration of the timeline, specifying the number of duplicates you want to create.

One point worth noting though, is the ability to distinguish between the "hard" Copies option, and an Alias or Clone. In most cases, you'll want to select the Copies options, so each newly created region is a discrete copy in its own right – separate from the original region. With an Alias or Clone, though, Logic will always reference the original region – so, if you make any subsequent physical edits to the region (like trimming off the last bar, for example), all the other cloned regions will follow this edit.

Loops

As an alternative to using Copies and/or Aliases, consider using the region looping feature. You can apply a loop in a number of different ways: including the Inspector's Loop parameter, the keyboard short L, or by using the unique Loop Tool that should appear at the top right-hand edge of the region. A loop is indicated by a grayed-out version of the region extending to the right-hand side of the original part. By default, a loop will last for the full duration of the song, or until it hits another region on the timeline.

Looping is a great way of quickly extending a song and sketching out an arrangement without having to create multiple duplicate regions. As you'd expect, edits that you make to the original region will be reflected in the

Figure 5.4 Use the menu option Region > Repeat Regions to quickly make a series of duplicate copies of a selected region.

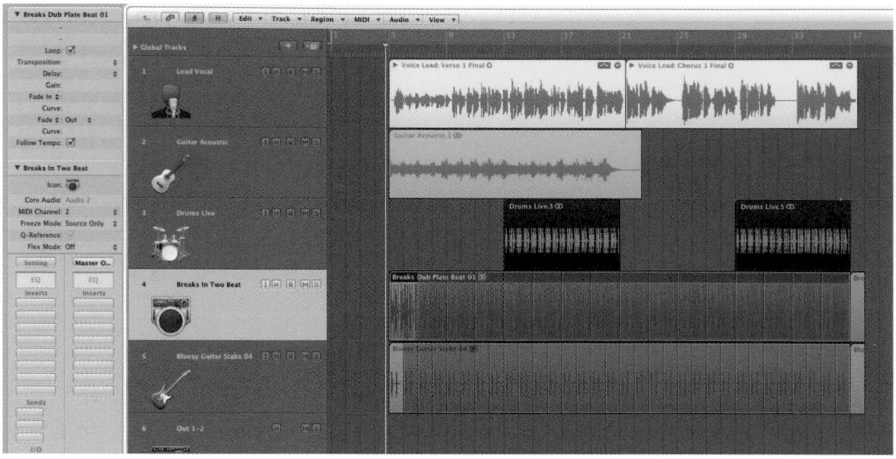

Figure 5.5
Region looping –
which is accessible
from the Inspector –
is a quick-and-easy
way of repeating a
region throughout
the duration of the
song (or until the
region hits another
object).

subsequent loops. This is much easier to keep track of than the region alias feature (thanks to the use of gray shading), and can even be put to use on some interesting creative treatments. One of the best examples of this is creating loops within your song that don't fall on exact bars – like a three-beat phrase, for example, or a five-beat motif. The result creates constant movement in the song, especially when played against a rigid 4/4 loop.

5.5 Precise Region Editing

Having established the broad structure of your song, it's now time to start to look at some of its finer details – from the precise placement of individual notes, for example, through to the task of "comping" between a number of different performances. Not surprisingly, this will start to push our use and understanding of Logic's editing features much further, utilizing a greater range of editing tools (like Flex Time or Quick Swipe Comping), as well as our ability to handle audio regions at much finer division than just bars and beats. The result, however, will be a significant improvement in the tightness of our performance, and the musicality of the track in general.

5.6 Snapping – Understanding Your Editing Grid

So far, the edits that we've made have been largely based on bar divisions – for example, duplicating two bars of drums throughout the duration of a song, or offsetting a bass part so that it starts on bar five. More refined editing through, requires a finer grid to work from – moving from divisions based on bars and beats, right down to individual ticks, or even sample frames! By editing at such a detailed resolution, your edits will be that much more precise, allowing you to accurately position an edit to the most detailed timing resolution the human ear can perceive.

In Logic, the current editing resolution is defined by the Snap setting and can be found in the top right-hand side of the Arrange area. Options include various different timing divisions (like Bars, Ticks, Frames, and Samples) and an intriguing "Smart" mode that intelligently adjusts the editing grid based on your current zoom resolution.

At the start of a new project, the Snap value defaults to the Smart setting. This is an intuitive way of working with Snap, and to a large extent, provides you with the freedom not to worry about any current Snap settings as you perform your edits. Dependent on the zoom level, and the size of the region, the Smart mode interprets the division you require – if you're working at a "macro" level, for example (moving regions around to form the structure of your track), Logic will stick to moving regions in bar increments. Zoom in closer, though (maybe to edit an individual note within a performance), and Logic will work at a correspondingly finer grid – whether it's sub beats, for example, or individual frames.

Another important point to note in editing resolution is that Logic always attempts to keep the relative positions within bars, beats, and so on. Therefore, precise, nonstandard edits can remain relative to the "snapped" grid when moved or copied to other position in the project – a bass part, for example, that picks up 1/16th before the start of bar 17 retains this offset when moved a bar ahead in the track (even if the Snap value is set to Bar). Again, this allows you

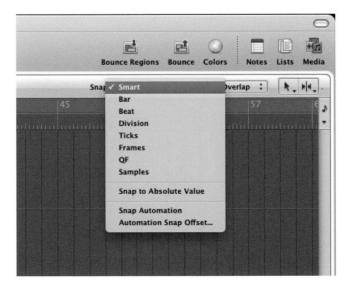

Figure 5.6 Logic features a variety of Snap settings, although the "intelligent" Smart option is best for day-to-day production activities.

not to worry about Snap settings and region offsets as you go about making your arrangement, although it's also worth remembering that just because a Snap setting is set to Bar, for example, doesn't always mean that the region will always be precisely placed on a bar.

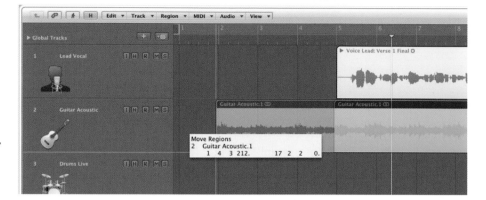

Figure 5.7 Logic tries to preserve any established "positional offsets" as you move a region around the Arrange area.

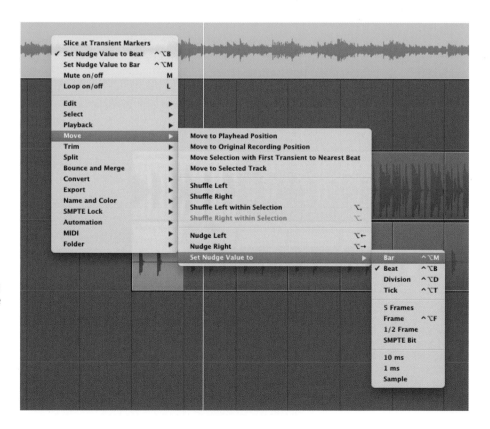

Figure 5.8 Using the Nudge feature is a great way of fine-tuning a region's position. Change the current Nudge value by opening up the contextual menu.

Going Finer – Bypassing Snap and Nudging Regions

Of course, there are times when you may wish to bypass the snap constraints, even when it's set to the intelligent Snap mode. This can be easily achieved by holding down Ctrl whenever you use a corresponding tool – like slicing a region, for example, or moving a region using the Pointer Tool. Another option is to use the Nudge option, which is specifically designed so that you can move audio region in small increments. A region can be nudged by selecting it, and then its position can be adjusted using either the left or right cursor keys combined with the Alt key.

As a quick way of adjusting the current Nudge value, try Ctrl-clicking on the region and selecting Move > Set Nudge Value to from the drop-down contextual menu. By using this feature, you can quickly set the Nudge value to the most appropriate setting – whether it's as fine as a Sample, for example, or even Bar and Beat settings to maneuver regions about on the timeline using a keyboard rather than a mouse.

5.7 Resizing and Cutting Regions

Having understood how to work with various different snap resolutions, let's start to explore some of the finer aspects of trimming, cutting, and resizing regions.

Remember that all the edits that you make on the Arrange area are nondestructive, and that once you cut a region, it's always possible to resize the region and restore the original information. This makes it easy to audition possible edits, being confident that it's always possible to return to the original recording at any point in the production process.

Cutting a Region with the Scissors Tool

One of the most immediate ways of editing is to use the Scissors Tool to divide one or more regions into two. An extra dimension to the Scissors Tool is that you can scrub across the timeline to hear where you are about to place an edit. Once you've clicked on the region/s to be edited using the Scissors Tool, simply hold the mouse button down, move the mouse back and forth to "scrub" the edit, and then release the mouse button when you're confident of having found the correct location.

Another useful addition to working with the Scissors Tool is the ability to quickly slice a region into a number of equally sized slices – like 1/8th or 1/16th, for example. To cut a region in this way, simply place an edit at the start of the region that equals the same size as the subsequent slices (like 1/16th). However, before you release the mouse button, hold down the Alt key, noting the small

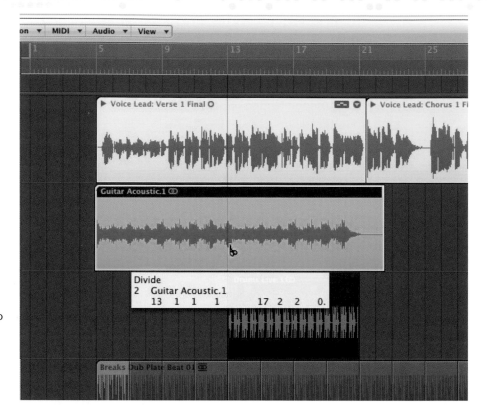

Figure 5.9 Use the Scissors Tool to divide a region in two. Remember, you can scrub the edit point to fine-tune the position of the cut.

Figure 5.10 Holding down Alt as you make a cut allows you to create a series of equally sized divisions. This is a useful technique for glitch and stutter editing.

"+" icon that should be added to the Scissors Tool. On release, the region should now be sliced into a series regions all of equal length. Try this technique as means of creating "glitch and stutter" type edits, especially using 1/16th and 1/32nd segments of a performance strategically rearranged and looped accordingly.

Topping-and-Tailing

Another important editing technique with audio regions is the so-called topping-and-tailing. This ensures that there are accurate and clean edits at the front and end of the audio passages, thus reducing any unnecessary noise and other unwanted artifacts. It also allows you to set a precise length for a region, which is vital in situations where you want it to loop effectively.

Regions can be manipulated easily and trimmed to fit by simply calling up the Resize Tool, which appears when you move the pointer to the bottom left- or right-hand extremes of any region. The pointer changes to an icon similar to that of a squared bracket with arrows on either side. With this tool enabled, simply drag the region to the desired length. As previously mentioned, this is as useful way of restoring an audio region to its original form if you've been slightly over zealous with your editing!

Using the Marquee Tool

A useful alternative to the Scissors Tool is Logic's Marquee Tool. Rather than cutting a region in two, the Marquee Tool allows you to select audio events within one or more regions, and then either delete the offending notes, or indeed, create a new region based on the Marquee's boundaries. Overall, the Marquee Tool works best as a means of time-slipping notes, especially on a multitracked recording of drums – first, by selecting the offending hit in question using the Marquee Tool, separating the selection into a new region, and then moving the region/s accordingly.

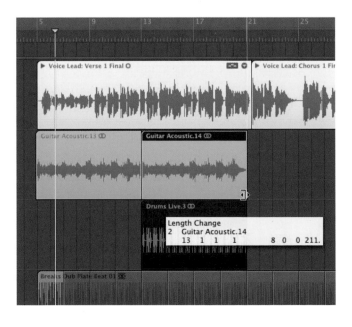

Figure 5.11
Moving to the edge of a region will transform the Pointer Tool into the Resize tool, allowing you to adjust the respective size of the region.

Once you've selected the Marquee Tool, simply drag-enclose the area you want to select – whether it's a few notes within a region, for example, or a portion of time across multiple regions. You can add to or remove from the selection at any point by Shift-clicking on your selection – adding in another track, for example, or increasing the size of the selection. Remember that any positional information will be set using the current Snap setting, so if you need your edits to be particularly precise, you'll either need to zoom in on your edit, or use the control modifier to temporarily disable any snapping.

Once selected, you can simply hit the Delete key to remove the offending selection – leaving two regions (and an empty space) on either side of an edit. Alternatively, press Mute (M) to silence the selection but leave the region in place. To divide the selection into a separate region, simply Ctrl-click on the Marquee selection, and then Logic will divide the original region into three new regions.

Snap to Transients

A useful extension of the Marquee functionality is the ability to extend the selection to transient hits within the region. This is a particularly good way of

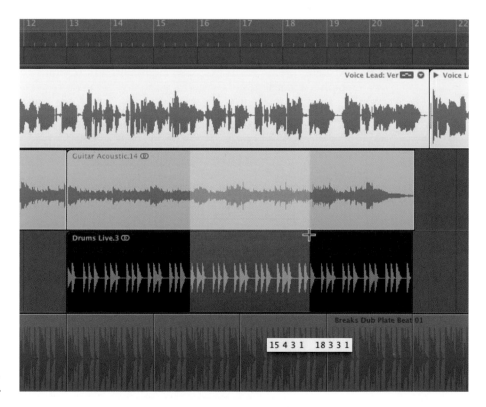

Figure 5.12 Use the Marquee Tool to enclose a selection across one or more regions – this can then be separated into a new region, deleted, or muted.

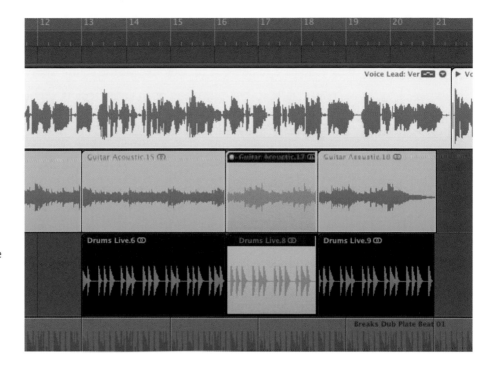

Figure 5.13 The Marquee Tool is a useful way of muting audio events within a region – simply enclose the area and press M.

editing drums, allowing you to accurately define a Marquee selection to hits within the region – this would allow you to neatly separate out a two-bar drum loop, for example, precisely slicing out an individual drum hit. Start by making the Marquee selection over the broad area you want to work with. Next, move the start or end points of the selection using either the left and right cursor keys to move the out-point of the enclosure, or the left and right cursor keys with the Shift key held down to reposition the in-point of the enclosure.

5.8 Fades and Crossfades

Fades and crossfades are an important part of audio editing on many different levels – whether it's just top-and-tailing an audio file, for example, or splicing together two adjacent audio files without creating any unnecessary clicks, or transitional discontinuities, between the two regions. In Logic, there're three different ways to create fades and crossfades, each of which has its own strengths and weaknesses.

Using the Crossfade Tool to Create Fades

Probably the most immediate way of applying a fade is to use the dedicated Crossfade Tool, which can be used to create both fades and crossfades. The advantage here is that the tool works in a graphical way, with the length of the fade established by the proportion of the region you enclose using the Crossfade Tool. This makes it easy to establish the fade in relation to the waveform – maybe fading a region out, for example, as the sound decays away.

To create a fade at the start or end of an audio region, simply drag the Crossfade Tool over the appropriate segment you want to fade. Logic will then update the region, displaying a shaded area to indicate the length of the fade. To change the fade at this point, simply drag the small vertical white line accordingly, moving it backward or forward to lengthen or shorten the fade. Clicking on the fade itself allows you to adjust the relative curve of the fade – moving between a linear fade (the default) and the exponential and logarithmic curves.

Where two regions are "butt edited" up against one another, you might want to consider placing a crossfade. A crossfade is designed to smooth out potential discrepancies between the two regions, so that you're left with a smooth transition between the two components. For example, even if the two regions are related in some way, the process of butting one up against the other will result in a small click as the waveform jumps from one position to another. In worse case scenarios, it might be that the two takes are radically different from one another, possibly necessitating a longer crossfade so that the edit isn't noticeable.

You can create a crossfade by dragging the Crossfade Tool across two adjoining regions, with the respective length of the crossfade dictated by the size of

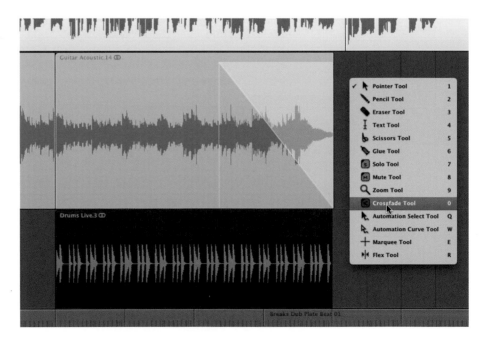

Figure 5.14 Use the Crossfade Tool as a simple, graphical way of fading a region in or out.

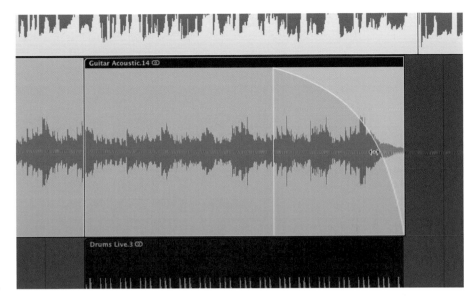

Figure 5.15 With the Crossfade Tool active, click on the white fade line to change its curvature or length accordingly.

the enclosure you make. Of course, in the case of a discrete butt edit, it might be that only a small crossfade is required, although this is usually best placed just ahead of a beat so as to not to damage any transient information. Where the discontinuities are greater, though (for example, maybe the adjoining notes have slightly different tuning), you'll need to consider using a longer crossfade.

Changing the Drag Mode to Create Fades

Along with the various Snap modes, Logic also includes several different Drag options, which can be found in the top right-hand corner of the Arrange area. The Drag mode defines how two regions interact when they are dragged over one another.

By default, regions are set to the Overlap drag mode, meaning that the regions can sit on top of one another, but the latter part will always take priority and mute the region that lies underneath it. This is a logical starting place for editing, especially when you might want to slip back and forth the crossover points between the two regions so that they fit appropriately. No Overlap, however, simply means that the first region will be edited in size to fit.

As an alterative to overlapping regions, Logic can also employ an automatic crossfade tool within the Drag menu, known as the X-fade. In X-Mode, the process of moving or resizing a part over another audio region will create a correspondingly sized crossfade. Therefore, by careful use of the region resize tool, you can quickly produce perfect crossfade edits between multiple regions – initially "butting" the edit to a bar, and then, resizing the inserted region/s back to create the crossfade.

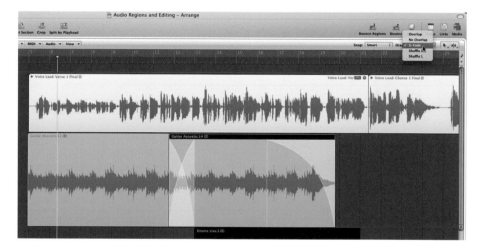

Figure 5.16 The drag mode defines how Logic will behave when two regions are placed on top of one another.

Figure 5.17 With the X-fade mode engaged, regions that overlap (either by resizing, or being moved) will be crossfaded accordingly.

Using the Fade Parameter to Create Fades

The final method for applying fades is to use the Fade feature as part of the Inspector's Region Parameters. Given the intuitive methods of applying fades that we've already seen, you'd be forgiven for wondering why you'd bother using the Inspector method. The first significant advantage with this approach, though, is the ability to create incredibly short fades without zooming in to the audio region/s in question. Second, it also provides a convenient solution where you need to apply fades to multiple regions at once – something that is impossible to achieve using the Crossfade Tool.

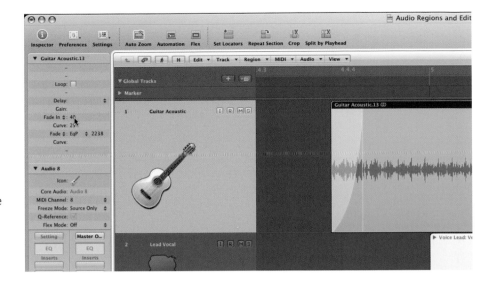

Figure 5.18 The Inspector offers a precise way of entering in fade information for a given region.

5.9 Speed Fades

As an extension to the fade parameter found in the Inspector, it's also possible to create a Speed Fade. Technically, a Speed Fade isn't something you'd want to apply in conventional audio-editing activities, although as a creative effect, it's a joy to behold! In theory, a Speed Fade approximates how a DJ might "spin in" a record, as well as slowing it down at the end to create a dramatic tail-off. Therefore, rather than modifying the amplitude of a region, a Speed Fade modifies its speed – pitching the region up across a "fade in," while slowing it down over a "fade out."

You can create a Speed Fade in much the same way as you create an existing fade. It probably makes most sense, therefore, to start by creating a traditional fade out – using the Crossfade Tool – over the area you want to Speed Fade. Once this has been established, turn to the Inspector and change the fade mode from Fade In or Fade Out, to Speed Up or Speed Down, respectively. Logic should now apply the fade, but in the form of a Speed Fade. Any subsequent changes to the length of the fade, therefore, will now change the duration of the speed-changing effect.

5.10 Quick Swipe Comping

In Chapter 4, we introduced the notion of cycle recording and Logic's ability to keep these multiple performances – or takes – as part of a Take Folder. This is certainly an effective way of managing multiple recordings of the same performance, reducing potential clutter on screen, and allowing you to group related recordings in a logical way. However, once you've recorded all these different

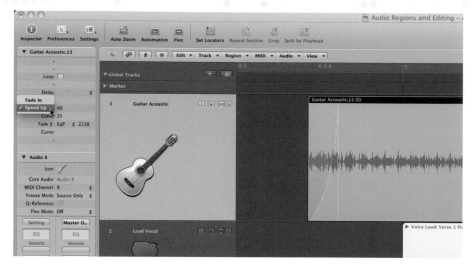

Figure 5.19 Use the Inspector to change the fade type – this allows you to create a distinctive Speed Up or Speed Down fade.

takes, you then need to decide which takes you want to use, potentially switching between several different takes as part of the same finished "performance."

Technically speaking, the process of editing between different takes is known as comping, and Logic includes a dedicated feature specifically for this task – Quick Swipe Comping. Using this feature, you can quickly audition different takes, as well as dynamically switching between them without having to use complicated editing tools and crossfades. You also have the ability to create several different comps that allow you to experiment with different treatments of the same musical part without damaging any existing comp decision you might have already made.

5.11 Expanding the Take Folder and Creating Basic Comps

You can expand a Take Folder to see its contents by clicking on the small arrow in the top left-hand corner. You should now be able to see the various takes, alongside shading to illustrate which take is currently being used. At this stage, you could audition each take – either by using the small drop-down menu in the top right-hand corner of the Take Folder or by simply clicking on each respective take.

The idea of Quick Swipe Comping is that you can quickly move between the different takes – "swiping" the word, phase, or series of notes that you want to include. Adapting the quick swipe edit points between the different takes is as simple as dragging over the respective take to change the amount of highlighted audio. Note how changes on one take are reflected in another (only one take can play at any point in time), and how the top line displays the current edit in its complete state.

Comps that you have made can be named using the drop-down Quick Swipe Comping menu. Note that it's also possible to Duplicate and Delete multiple different Comps, allowing you to try out several different editing strategies, and then audition the relative merits of each, by switching between the different completed Comps (which, again, are selectable from the Quick Swipe Comping drop-down menu).

Figure 5.20
Once you've opened your Take Folder, you can quickly start to piece together a preferred Comp between the various takes.

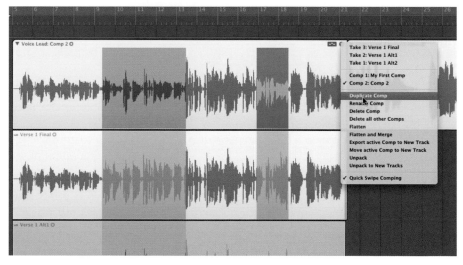

Figure 5.21
Comps can be named, duplicated, and deleted, allowing you to explore multiple different versions of the same line.

5.12 Time Slipping a Quick Swipe Comp

So far, we've seen how we can use the Quick Swipe Comping system as a means of switching between different takes of the same musical part. However, what we haven't yet addressed is the need to slip potential edit decisions that you've made in time – of course, you can always move the whole Take Folder, but what if your concerns are with only one or two notes within one of the takes? If you attempt to move a region at this stage, you won't get very far – either because of the inability to use the Scissors Tool, or that any attempt to move the region results in the Comp itself being changed.

The solution to this problem is the ability to switch the Take Folder's Quick Swipe Comping functionality on or off accordingly. You can see whether Quick Swipe Comping is enabled by looking at the top right-hand corner of the Take Folder, just to the left of the Take Folder's drop-down menu. By disabling the Quick Swipe Comping, you'll find that Logic's original toolset – namely, the Scissors Tool and Marquee Tool – now work as expected. Given that the broad comping is finalized (this time slipping is much easier to achieve after you've

Figure 5.22
By disabling the Take Folder's Quick Swipe functionality (by clicking on the small icon on the top right-hand corner of the Take Folder), you can use the standard editing tools, particularly with respect to time-slipping elements of the performance.

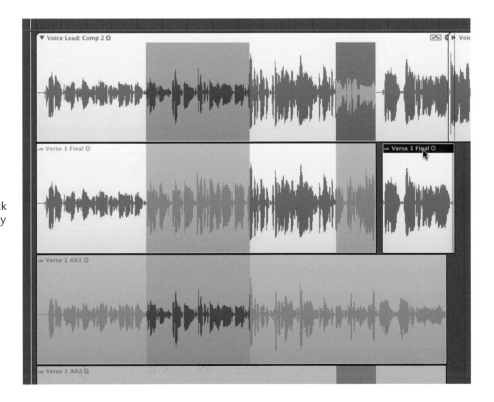

swiped the takes!), you can go about slicing out the offending notes and then repositioning them.

As we'll see later on, though, this isn't the only solution for time-slipping material within a comp.

5.13 Flattening or Merging a Comp

Finished, or even partly finished, comps can be handled in a number of different ways. First of these, is the option to flatten the edits you've made – "unpacking" the Take Folder, and creating a single track with all the edits in place. Flattening has its advantages and disadvantages – on the negative side, you can't return to the edits in such an intuitive way as the Quick Swipe Comping feature, but on the positive side, you'll gain a "clean" edit of your chosen comp, complete with editable crossfades within each edit point.

You can choose to flatten your edits through the Take Folder menu, with the resultant output displayed as a series of regions with crossfades between them. Flattened in this way, you can easily carry out any timing modifications (using an alternative approach to what we've seen previously), as well as adjusting the default crossfade times using the same processes that we've explored earlier on in the chapter.

A more destructive version of the Flatten function is the Flatten and Merge option. Using Flatten and Merge will create a new file, complete with the edit points and crossfades. If you're completely happy with your edits, Flatten and Merge could be an appropriate choice, especially if you want a more transportable version of the track to import into another DAW. Unless you choose to specifically delete the original takes, you'll always have the option of reimporting the files from the Audio Bin into the Arrange area should you decide to revisit what you've done at a later point.

Whether you're carrying out a Quick Swipe Comp or Flattening a Comp Logic will work with the current default Crossfade Time and Curve settings established under the General tab of Logic Pro > Preferences > Audio. In most cases,

Figure 5.23
Flattening your comp will create a single track with all the various takes edited accordingly.

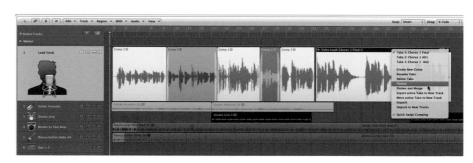

Figure 5.24
Flatten and
Merge is a more
destructive way
of finalizing your
comp – with a
completely new
audio file created
from your editing
decisions.

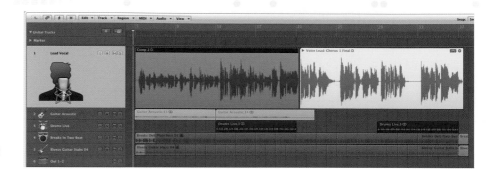

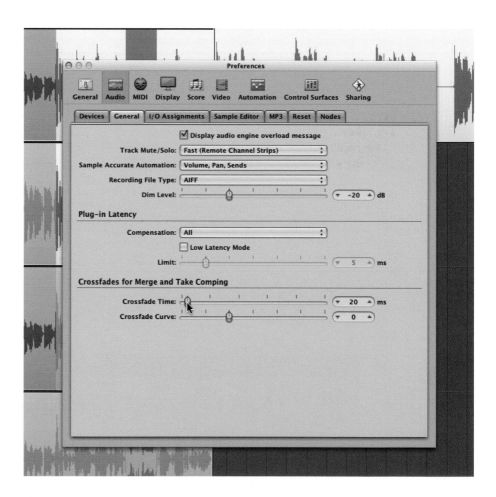

Figure 5.25
Use the General
tab, found under
Logic Pro >
Preferences >
Audio, to change
the default
crossfade time
for Quick Swipe
comping.

the default setting of 20 ms is sufficient, although for particularly taxing comps, it might be the case that this has to be raised or lowered accordingly. Arguably, though, such fade adjustments are best applied to a flattened comp and this preference left as is.

Unpacking a Comp

As an alternative to flattening your edits, it's also possible to unpack the Take Folder, placing all the individual takes, as well as the various comps (complete with crossfades), onto separate tracks. Unpacking a Take Folder is a messy option in terms of the Arrange area and the amount of tracks it can potentially use up, but it does offer the best solution in terms of complete audio flexibility. For example, an unpacked take can be easily moved to new tracks – maybe using some outtakes from a main vocal as backing vocals. Alternatively, it might also be that the remit of editing goes above and beyond what can be achieved in Quick Swipe Comping alone – like a vocal requiring pitch correction in Melodyne, for example, or elements of time compression and expansion.

As the reverse of unpacking Take Folders, you can also use the menu function Region > Folder > pack Take Folders to combine several existing audio files into the same folder. The option to pack Take Folders potentially solves the previously suggested problem of adjusting timing issues – simply unpack the folder, move the required take, and then repack accordingly. The feature can also be an interesting way of creatively switching between a number of concurrent drum loops – packing them into a shared Take Folder and then using the Quick Swipe Comping to create complicated jumps between the loops.

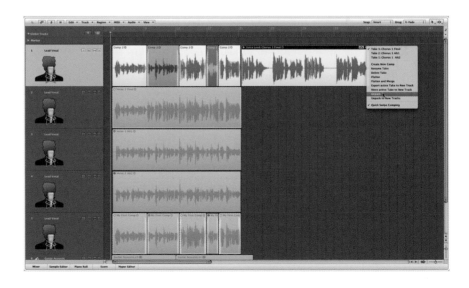

Figure 5.26
Unpacking a Take Folder will explode all the original takes onto separate tracks, as well as flattening any comps you've made along the way.

5.14 Flex Time Editing

So far in our exploration of production process in Logic, we've looked at the "conventional" tools and techniques of audio editing – namely, the ability to slice, resize, and reposition audio regions. However, in the past 5 years or so, musicians and producers have grown increasingly used to having a greater degree of "elasticity" with their audio – either using specific third-party applications like Melodyne, for example, or an increasing amount of tempo and time-adjustable functionality directly embedded into a DAW of choice. Therefore, having already seen some of the tempo-elastic possibilities of Apple Loops, Flex Time editing transforms the potential flexibility of audio regions within your project.

Why Use Flex Time?

As the name suggests, the principle use of Flex Time is to modify the timing properties of an audio region and the "events" contained within it. Of course, using conventional editing techniques, we can always use region slicing to create such timing moves, but this can often be a laborious task necessitating multiple slices and complicated crossfading to make an edit sound transparent enough to the listener's ear. What's next to impossible to achieve using standard editing techniques, though, is the ability to change a note's duration, whether it's shortening or lengthening a note to optimize its phrasing.

The first application of Flex Time, therefore, is the ability to move notes backward and forward within a given region – achieving everything from the ability to "slip" a note so that it plays more in time with events in other audio regions, through to the ability to completely rephrase a melody line (including, as we've seen, the duration of the notes) so that each note falls in a different position in the bar. As the audio is "intelligently" time stretched (using the currently selected Flex Time mode) no additional editing is required, and the transition between the edited and unedited notes remains smooth and glitch-free.

Another important application of Flex Time editing is the ability to quantize audio events within a region – making a bass play exactly in time with a drum loop, for example, or changing its groove from a straight 16th feel to a bassline with a subtle "swing" feel. Again, this is something that's possible to achieve with conventional editing but can be achieved in Flex Time in a matter of seconds.

5.15 Flex Time View and Flex Time Modes

To start working with Flex Time in Logic, you'll need to engage the Flex View mode (View > Flex View). Once the Flex View mode has been engaged, you see some extra information in each of the track headers, allowing you to select

Figure 5.27
To begin working with Flex Time, you'll need to engage the Flex View mode (View > Flex View).

a unique Flex mode for each track in your project. Ideally, you should try to match the Flex Time mode to the type of track being handled. The modes are largely self-explanatory, but let's take a look at their principle application in music production.

Slicing

The Slicing mode is the obvious choice to use with drums and percussion sounds, and makes a point of preserving transient details in the original recordings. Behind the scenes, the audio region is "flexed" using slices, using much the same process as we've seen with manual editing.

Rhythmic

The Rhythmic mode is a good "catch all" solution for a range of audio files that contain both pitched and percussion elements – this might include a drum loop with some in-built bass parts, for example, or a rhythm guitar. Some "stretching" artifacts might be evident under closer scrutiny (or extreme settings), either as part of the body of the sound or the transient details.

Monophonic

If you working with a "melodic" instrument that doesn't play chords – like bass, vocals, or lead guitar – then opt for the Monophonic Flex mode. This provides a good balance between transparency and CPU usage.

Polyphonic

The Polyphonic mode is arguably the most CPU-intensive Flex mode, and should be reserved for polyphonic instruments, or indeed, complex submixes, that require high-quality time stretching with any unnecessary artifacts kept to a minimum. Remember that you can always bounce the finished Flex Time edit (for more information, see Section 5.19).

Tempophone

Definitely an inappropriate choice for smooth "flexing," Tempophone is a Flex mode deliberately designed to sound obtrusive and noticeably modified. Use Tempophone when you want to use Flex Time almost like an effect, with the sound appearing increasing grainy and stretched as you move increasingly further away from its original duration.

Speed

As the name suggests, the Speed Flex mode modifies the speed of an audio event to account for changes in time. Compacting the placement of notes, for example, will speed up the playback, whereas stretching out notes will slow it down. This is much the same as modifying the pitch and tempo of track using a turntable's varispeed control, or indeed, modifying the playback speed of tape. The Speed Flex mode, therefore, is best reserved as a creative effect, where you want audio to sound deliberately repitched accordingly.

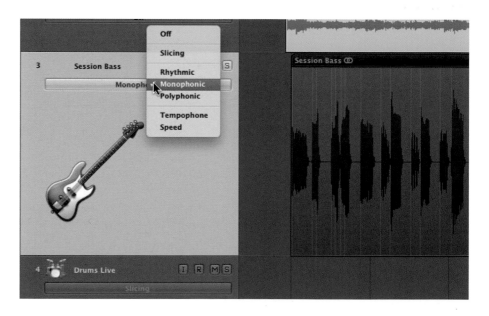

Figure 5.28
A variety of different Flex modes allow you to optimize the time-correction algorithms for the type of audio material you're working with.

Besides engaging the Flex mode for the track, it's also worth to check that the Flex functionality has been engaged as part Inspector's Region Parameters box. In most cases, this will already be activated, but occasionally a region can have its Flex functionality removed.

5.16 Transient Markers and Flex Markers

Flex Time makes use of two types of makers – Transient Markers and Flex Markers – to carry out the "flexing" process. Transient Makers are first used to divide up the audio events within a region – creating markers for each word on a vocal take, or the individual hits within a drum performance. Flex Markers, however, identify the notes you wish to modify and how the notes are handled on either side of the Flex edit. Although this sounds complicated at first, their use and application soon becomes apparent once you start using Flex Time.

Creating Transient Markers

Whenever you engage a Flex Time mode, Logic will go through the regions on the selected track and assign a series of Transient Markers. This is a relatively quick process, and once finished, you'll see a series of small gray hairlines to indicate their placement. Note that with Take Folders, though, the markers aren't evident on the "top" level but can be viewed and modified once the Take Folder has been expanded.

Flexing Multiple Audio Events

With a series of Transient Markers placed, you can now start to explore various applications of the Flex Time technology. Probably the first and most immediate application is to resize a region and its contents accordingly. Previously, of course, the act of resizing a region would result in the "edit window" being resized, rather than modifying any of its contents. In Flex Time mode, though, the events will be expanded or contracted accordingly. This make it easy, for example, to conform a drum loop imported in at one tempo to your current project's tempo, simply be resizing to the corresponding region end point.

Note that the expansion or contraction of any audio information within the region is always color-coded accordingly. Audio events that have been contracted will take on a green hue, whereas those that have been expanded will take on an orange shade. Where the stretch is too extreme, the shading changes to a red color, suggesting you try an alternative solution to the problem at hand!

Figure 5.29
Once a region has been placed in Flex mode, any attempt to resize it will result in the audio being expanded or contracted accordingly.

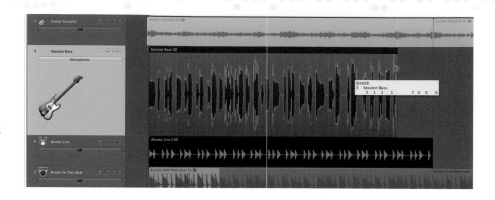

Figure 5.30
A single Flex Marker will expand or contract the audio events on either side of it, as indicated by the green and orange shading.

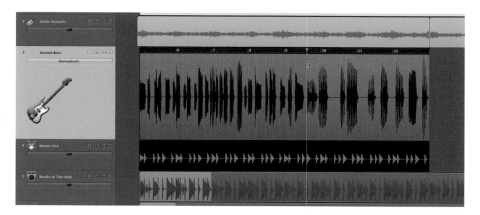

Besides changing the region as a whole, Flex can also modify notes within it using a series of one or more Flex Markers. The art in understanding and using the Flex Marker system lies in the two contrasting ways Flex Markers can be placed, as indicated by the changing appearance of the Flex Marker Tool that appears as you float over the region in question.

By clicking in the upper half of our waveform display, we can create a single Flex Marker, which is set to the nearest Transient Marker and effectively acts as an "anchor" to the flexing process. By moving the Flex Marker in either direction, audio will be expanded or contracted on either side of the marker, again using green and orange color coding to indicate the qualities of expansion and contraction taking place.

Flexing Single Notes

A far more musical way of using Flex Time is to use three adjacent Flex Markers to move a single note backward or forward accordingly. Although the process of creating multiple Flex Markers might sound complicated, it's easy enough to execute. With the Flex Time view active, move the pointer to the lower half of the region and note how the icon changes from one marker to three. Now, click on the note you want to move and note how three markers are created – one preceding your editing point, one on your edit point, and one succeeding your edit point. Moving the central Flex Marker moves the desired note position but preserves the timing on either side of the edit.

Another interesting way you can apply Flex Time is to tighten notes between adjacent regions – maybe aligning a note on a bass guitar, for example, to precisely play in time with a hit on the kick drum. First, you'll need to ensure that both tracks have been assigned a Flex mode, and the track that you want to

Figure 5.31
Placing three Flex Markers allows you to move a note's position without affecting the timing of notes on either side of it.

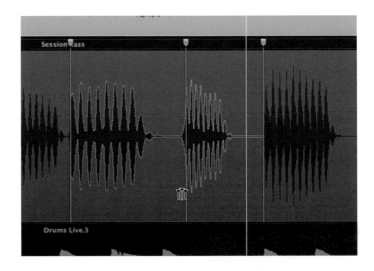

Figure 5.32
By moving to an adjacent tracklane, you can snap a note so that it aligns with another part.

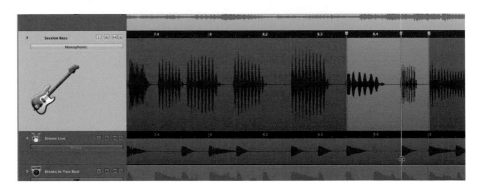

lock the note (or notes) to has been shuffled up or down the tracklist so that the two sit adjacent to one another. Again, we start to move our selected note using the "three marker" icon found in the lower half of the waveform display. This time, however, drag the mouse up or down to the adjacent track and note how the Flex Time marker snaps to transients on this track (the positioning line should also change to yellow to indicate this is happening).

Deleting Flex Markers

In keeping with the nondestructive ethos of editing audio in Logic, it's also worth noting that Flex Time edits can be undone at any point in the production process, returning the audio region back to its original state. One gradual way of achieving this is to delete individual Flex Markers, thereby removing any unwanted Flex Time edits, but retaining elements that you are happy with. To delete an unwanted Flex Marker, therefore, Ctrl-click on the Marker and select Delete Flex Marker from the small drop-down menu.

When multiple Flex Markers have been assigned and you want them all to be removed, you can always Ctrl-click anywhere else on the region and select Delete All Flex Markers. This will restore the region to its original state, although subsequent regions on the track will retain any Flex Time edits that you have made. Therefore, to disable any Flex Time editing across the whole track, you can always switch the Flex mode back to its "off" setting to completely disable Flex Time editing.

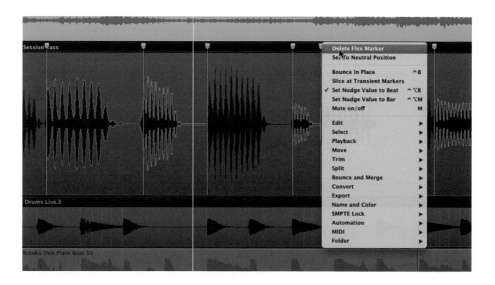

Figure 5.33
Deleting a Flex Marker by Ctrl-clicking on it and selecting Delete Flex Marker.

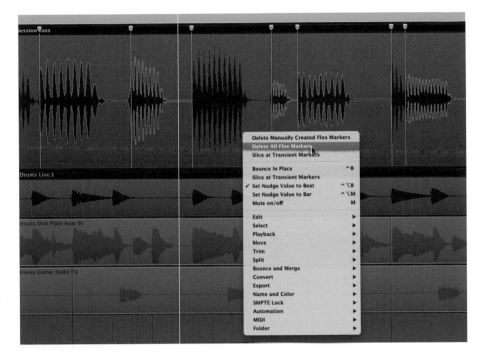

Figure 5.34
As well as deleting individual Flex Markers, you can also delete a series of Flex Markers assigned to a region.

5.17 Quantizing Audio with Flex Time

One of the standout features of Flex Time has to be the ability to quantize the events within an audio region. Quantizing is the process of "tightening" a performance so that notes within fall more closely to the divisions of a bar – something that up until the introduction of Flex Time had largely been the preserve of MIDI editing (covered in detail in Section 6.8).

To quantize the events in an audio region, you first need to select the region/s you want to quantize, and then pick an accompanying quantize setting from the Inspector's Region Parameters box. There're various settings to choose from, although in most situations you'll want to select the 1/16 setting, which places each note to the closest 16th (or semiquaver) division of the bar. Once selected, you'll notice that each note within the region has been moved slightly (unless the part was played exactly to the beat), as indicated by the orange and green shading.

From this basic quantize setting, we can then adjust various settings to change the feel and strength of the quantizing applied. For the sake of brevity, though, turn to Section 6.8 to understand fully what can be achieved here.

Groove Templates between Regions

A handy extension of the basic audio-quantizing facilities using Flex Time is the ability to copy the groove of one audio region to another. In many ways, this can be a more suitable way of improving the timing between different audio regions – allowing two different parts to play tightly against one another, but not losing the feel of the song to a rigid 1/16th grid.

The first part of the process is to extract a groove template from a suitable source region – in other words, the part you consider to be "in time." To extract the groove, the track will need to be placed into an appropriate Flex mode (slicing is probably best) so as to create a series of Transient Markers. Once this has been done, highlight the region, and then select Make Groove Template from the region Inspector's quantize menu. Once the groove has been created, you can choose to take the track out of its current Flex Time, assuming you have no need for it to be "flexed" in any way.

Applying the newly created Groove Template is much the same as the example of audio quantizing we've already explored earlier on in the Flex Time editing

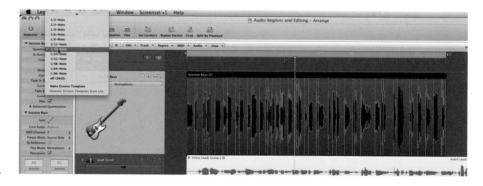

Figure 5.35
Flex Time allows you to quantize the performance of an audio region – shifting each respective note so that it precisely aligns to a rigid tempo grid.

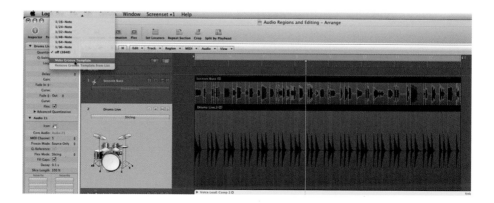

Figure 5.36
Using the Region Parameter drop-down menu, you can add the "groove" of an existing region to your palette of quantize options.

Figure 5.37
Having created our new groove, we can then apply it to another region in our song. The two parts should now play together perfectly.

section. In this case, however, you'll need to pick the particular Groove setting (rather than 1/16th, for example), which should appear at the bottom of the Quantize drop-down menu.

5.18 Separating Regions Based on Transient Markers

One final feature made possible by Flex Time is the ability to divide a single region into multiple regions based on Transient Markers within it. In the case of a recording of a drum kit, for example, this could allow you to create individual regions for each drum hit. Given the flexibility of Flex Time's slicing mode (which achieves similar functionality in a more transparent way), the wisdom of applying this is somewhat dubious, although you might find some benefits in having regions separated in this way. One useful application, though, would be to divide a recording up into a series of "one shot" audio files ready to be imported into a software sampler like the EXS24.

You can slice a region that has been placed into a Flex Time mode by Ctrl-clicking on the region and selecting Slice at Transient Markers from the drop-down menu. Of course, once sliced, you can always take the track out of Flex Time mode to avoid any unwanted CPU activity. Should you want the regions

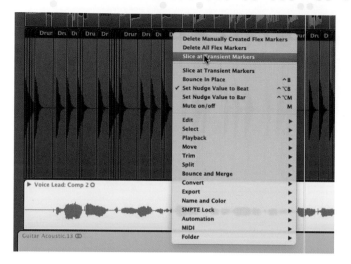

Figure 5.38
Once Transient Markers have been placed, you can separate a single region into a series of regions using the Slice at Transient Markers feature.

to become individual sound files in their own right, then highlight them, and select Audio > Convert Region to New Audio Files.

5.19 Freezing a Flex Time Track to Save CPU

Make no mistake about it – a heavy use of Flex Time editing can have a big hit on your CPU resources, especially if you use some of the more CPU-intensive Flex modes like the Polyphonic option. If you feel happy about the Flex Time edits you've made, therefore, you might want to consider "freezing" the track. Freezing is a common process in Logic (also covered in Chapter 8, Knowledgebase 3, Saving CPU: Freezing and More...) whereby an "off-line" rendered version of the track is created (complete with any plug-in settings or Flex Time adjustments), which is then streamed from disk in much the same way as an audio file. Although frozen, the track looks exactly the same – even retaining its Flex Markers – although any subsequent changes will result in the track needing to be rerendered, which can take a few seconds to complete.

To freeze a track, you'll need to click on the Freeze icon as part of the track header. If you can't see this, select View > Configure Track Header, and make sure the Freeze icon is checked accordingly. Before you initiate the freeze, another important point to check is the current Freeze mode, which can be established as part of the Track Parameters box, just below the Region Parameters box in the Inspector. Ideally, it should be set to Source Only, so that only the Flex Time edits are rendered in the form of an off-line file. By comparison, the Pre Fader option renders both the source information and plug-ins – which isn't so appropriate as we haven't got around to mixing our track yet!

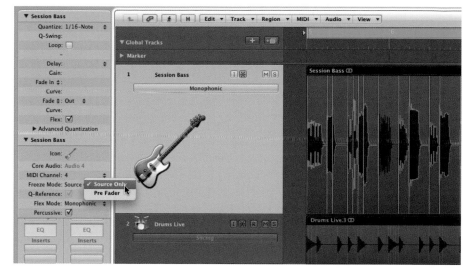

Figure 5.39
Using the Source
Only and Freeze
modes allows you
to render your Flex
Time edits as an
off-line file, making
a significant saving
on your CPU
resources!

5.20 Editing Multiple Tracks Using Edit Groups

Most of our edits so far have involved single-tracked instrumentation – that is to say, an instrument that's recorded to a single track in either mono or stereo. However, there are plenty of examples where you might want to record more than one track on a single "pass": for example, recording a drum kit; an orchestra using a combination of close microphones, room microphones, and a stereo pair; or a guitar amp using an array of different microphones placed around the cabinet. In all of these examples, your edits will want to be achieved en masse, rather than having to individually slice each accompanying region every time you make an edit.

Creating Edit Groups

Logic's Grouping function allows you to link together tracks both in respect to editing and mixing. In Chapter 8, we're going to be looking more closely at the applications of groups in mixing, but for now, we'll concentrate solely on their use in editing.

You can assign one of more tracks into an edit group through the Mixer area, which also provides a current overview of the grouping status. Each channel strip on the Mixer page has its own Group Slot, which you should be able to see just below its "Audio 1" legend. Therefore, to assign a channel strip to group, simply click on this box and select a corresponding group from the drop-down menu. Note that you can also mass-enable several channels strips by Shift-clicking to augment your existing selection, and then selecting an appropriate group using one of the channels strip's group slots.

Whenever you place a channel strip into a group, Logic will open the Group Settings window, which allows you to define how each group works. The principal option we're interested in at this point is the Editing (Selection) and Phase-Locked Audio check boxes, as this will facilitate our edits being simultaneously carried out across all the regions.

Once the group has been established, you can go about performing your edits using any of the techniques we've described – using basic editing features like the Scissors and Marquee Tools to perform rudimentary edits, or stepping up to features like Flex Time, where your edits need to be more detailed. One point to remember, though, is the choice of "master" region you use to perform your primary edits. In the case of a drum kit, for example, the Overheads track makes most sense as this gives an accurate overview of all the notes played, rather than the individual notes that relate to just one part of the kit.

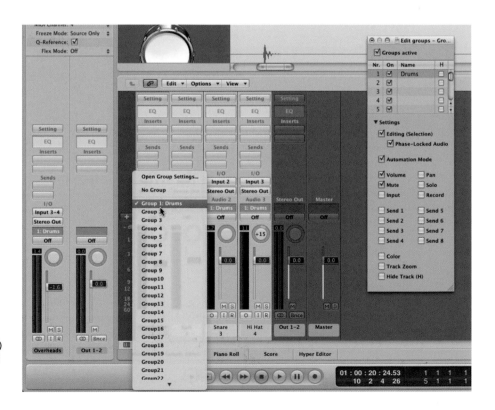

Figure 5.40
Create an edit group using the Mixer, ensuring you select the Editing (Selection) and Phase-Locked Audio check boxes.

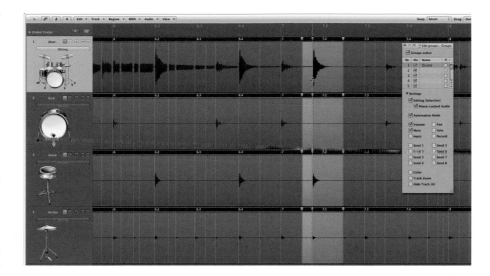

Figure 5.41
With the edit group active, all edits will be carried out across the associated regions. This is a powerful feature when used in combination with Flex Time editing.

Toggling Groups

Of course, once you've enabled a group, you're not restricted to all subsequent edits being carried out across all the regions. By toggling the groups – using the so-called Group Clutch key, which is assigned to cmd + G – you can disable the group functionality across all the active groups. This then allows you to carry out edits on individual tracks as appropriate, returning the group functionality (using the Group Clutch key) once the edits have been made. You can see whether or not groups are active by viewing the small group slot on the Mixer page – when the legend is yellow, the group is active, and when it's gray, the groups are currently disabled.

5.21 Drum Replacement

In addition to the use of edit groups and Flex Time to tighten a multitracked drum recording, you might want to turn to the Drum Replacement feature to either completely replace an individual drum sound within the kit recording, or to augment the existing sounds with a sample-based layer. Drum replacement is a common technique used in professional circles (although few actually admit it!), but has either previously required additional software – like Drumagog – or extensive use of the Audio to Score functionality (and plenty of laborious editing) to achieve in Logic.

Most drum replacement tends to focus on the kick and snare drum, as this contains the majority of the sound energy behind the kit, but equally because it's relatively easy to achieve. By comparison – the overheads would be next to impossible to replace, mainly because of the number of different sounds (kick, snare, hi-hat, toms, and so on) occurring at the same time. In short, therefore, stick to tracks that have been spot microphoned to a single part of the kit, rather than tracks with more ambient microphone placements.

To start the replacement process, select the track that you want to replace or double, and choose the menu option Track > Drum Replacement/Doubling. A dialog box allows you to refine the drum-replacement process using a few easy-to-use parameters. First, of course, you need to set the instrument type from the drop-down menu, which includes options for Kick, Snare, Tom, and Other. As you'd expect, the threshold control establishes the amount of triggers generated – if there's lots of background noise, for example, you might need to raise the threshold higher, otherwise pull it down to the point where all the drum hits have a yellow "trigger line" attached to them. You should be able to see all the triggers point on the accompanying regions.

The Drum Replacement feature works in conjunction with the EXS24 sampler, and as part of the replacement process, you should have noted that an instance of the EXS24 has been loaded, alongside MIDI trigger region, and a default sample setting. Therefore, once you've established the threshold, you might want to press the Prelisten button to hear the replacement in action (both the replacement part and the original audio track should play together solo'ed out from the mix), as well as stepping through the accompanying library settings to audition different EXS24 drum sounds. Once you're happy with the result, hit

Figure 5.42
Once you've chosen a track to work with, select Track > Drum Replacement/ Doubling to initiate the drum-replacement process.

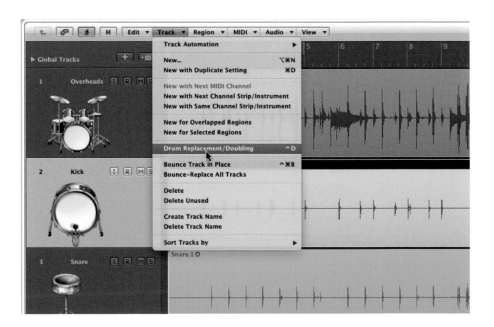

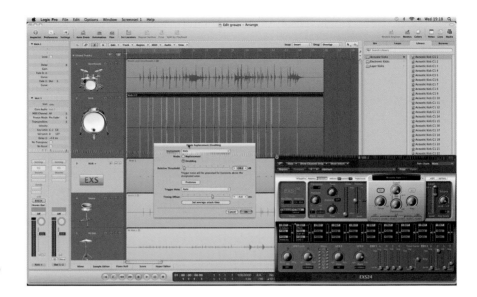

Figure 5.13
Using the Drum
Replacement
settings you can
fine-tune the
replacement
process, adjusting
the threshold
and the type of
sound you want to
replace.

the OK button to take you out of the Drum Replacement/Doubling dialog box, leaving the MIDI region and instance of the EXS24 in place.

Once you've created the basic trigger part, you can then spend plenty of time refining the results as you see fit. Of course, at any point you can change the selected EXS24 sample, especially as you get closer to the final mix and have a better idea of how the part should sound. It might also be relevant to take a closer look at the MIDI data that has been created – possibly deleting unwanted notes, for example, or changing the velocities to achieve the level of dynamics you require (we'll cover MIDI editing in more detail in Chapter 6). As the sample is devoid of any spill, there's also plenty of potential to add reverb, compression, and EQ, without the fear of what might happen to the rest of the kit.

5.22 Sample Editor Principles

Given the exhaustive editing possibilities within the Arrange area, you could be forgiven for wondering why it was necessary to have another "layer" of audio editing within Logic. The Sample Editor, though, adds some important additional features to working with audio in Logic universe, some of which might be less useful to your day-to-day workflow, but still essential in getting the most from Logic.

Rendered "Off-Line" Editing

The first important point to note about the Sample Editor Window is the fact that a large number of processes carried out within it are destructive. Rather than simply modifying playback information, therefore, many of the changes involve rewriting the data that is contained within an audio file – a gain change, for example, permanently modifies the amplitude of the data itself, rather than simply playing back the file slightly louder or quieter. Likewise, a timestretch will permanently render a longer or shorter version of the file accordingly. This technique of permanently changing an audio file is often referred to as an off-line edit.

Of course, actions taken in the Sample Editor have a degree of "undo" available to them, but it's important to remember that decisions made in the Sample Editor might irrevocably damage the audio files used in your project, not to mention the potential havoc caused by projects that share audio files among them. However, there're many examples of situations where you want to create a permanent change to the files, especially in situations where the audio file becomes the final "product" – like a sample ready to be imported into the EXS24 sampler, for example, or the final mix of a song ready to be burned onto an audio CD.

Therefore, as a good safety system, it's worth making copies of any audio regions you wish to transform using the Sample Editor. This can be done by

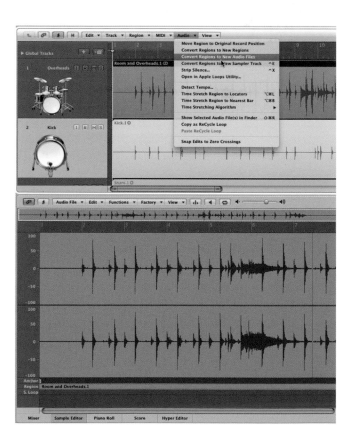

Figure 5.44
As the Sample Editor is a destructive process, you might want to consider creating a backup of your audio region before you perform any edits. One way of doing this is menu option Audio > Convert Regions to New Audio Files.

simply selecting the appropriate region and choosing Audio > Convert Regions to New Audio Files. A quick check in the Bin should indicate the creation of a new audio file, which you can modify, safe in the knowledge that the source files remains untouched. As an alternative, the Sample Editor also includes its own backup facility, found under the menu option Audio File > Create Backup.

Accuracy

Although it is possible to edit to a high degree of accuracy in the Arrange area, the Sample Editor remains one of the best places to edit audio information down to individual sample frames. Generally speaking, this happens very rarely. One possible example, though, is the ability to use the Sample Editor's pencil to tool to draw out small clicks that can appear in an audio file (especially when it has been poorly edited without any crossfades!). You can also use the Sample Editor to precisely align region start points so that they start precisely at the start of a note.

Additional Functionality

In addition to being able to perform off-line editing, there's also some useful "functional" additions that are primarily addressed in the Sample Editor – from the ability to edit Transient Markers, for example, through to creating and positioning so-called Anchor points. Again, these might not be so important to your day-to-day workflow in Logic, but they add to the repertoire of techniques at your disposal!

Figure 5.45
The Sample Editor allows you to view audio data at its finest resolution – even down to the ability "draw out" sample clicks and other unwanted information using the pencil tool.

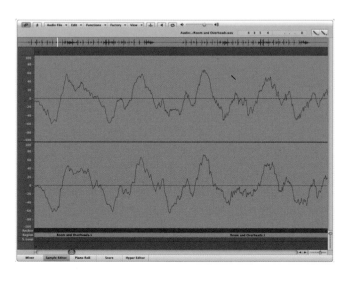

5.23 Sample Editor Applications

Although not exhaustive, we've compiled some of the best applications and uses of Sample Editor. There's plenty more to be found in this intriguing section of Logic, so it's worth exploring the Sample Editor furthermore as you grow more confident with your abilities.

Reversing Audio

This is one of my favorite applications of the Sample Editor, and it is one of the most rewarding ways of "twisting" a given sound – from reverse cymbals, through to backward guitar solos! Start by creating a duplicate audio file using the Audio > Convert Regions to New Audio Files command (otherwise, all occurrences will be reversed!). Once you've opened up the Sample Editor, highlight the area you want to reverse (usually the complete sound file) and then select Functions > Reverse.

To position a reversed region, take a look at the "Marking Anchor Points" feature outline below.

Accurate Region Resizing

Use the Sample Editor to change the in and out point of your regions, potentially negating any unwanted sounds that might precede the first note, for example, or stray sounds that occur after the last note. The current size of the region is indicated by the two markers at the bottom of the Sample Editor, with an "in" and "out" point accordingly. Changing this won't change any relative positional information, so even if you modify the start point (to remove some noise, for example), your region will still play perfectly in time.

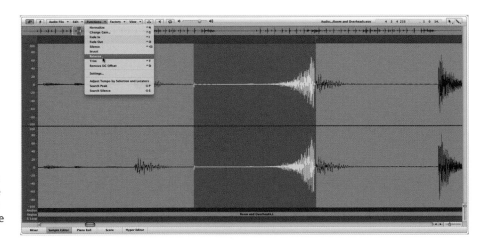

Figure 5.46
Use the Sample Editor's Reverse function to reverse a given sound – although remember this is a destructive off-line edit, so it's best to make a copy of the region first.

Marking Anchor Points

By default, all regions are positioned using their start points. However, it is possible to create an alternative anchor point positioned within the audio region. A good example of this in action is on reversed files, where the anchor point is probably best marked where the sound reaches its peak, rather than at the start of the region. You can place an anchor point by sliding the small orange marker at the bottom of the Sample Editor. Now, when you move the region in the Arrange area, you see the new anchor point being used to place the region, rather than the region's start.

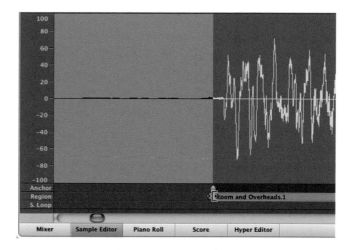

Figure 5.47
You can use the Sample Editor to accurately resize a region down to sample-level accuracy.

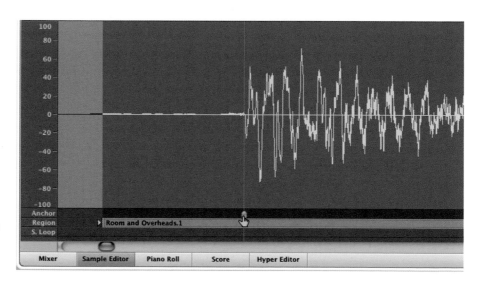

Figure 5.48
Anchor points are used as the "location reference point" for placing a region on the Arrange area.

Cleaning Up Unwanted Noise

Small unwanted noises in-between notes can be quickly removed using the Sample Editor. Simply drag-enclose the offending portion of audio and select either Silence or Change Gain from the Functions menu. So that there are no abrupt changes in the signal level, it might be prudent to opt for the Change Gain setting and pick a setting around –10 dB or so. This will attenuate the noise to the point that it is less noticeable but will not completely remove it. Again, remember these edits are destructive and will permanently change the source audio data!

It also worth noting the other "audio hygiene" options available under the Functions menu – from the ability to normalize an audio file (in others word, lifting its amplitude to hit 0 dBFS), through tools like the Invert function, which can be used to permanently change the phase of an audio file.

Audio to Score

One interesting feature of the Factory menu is the ability to transform an audio region into MIDI data. As you'd expect, the results that can be achieved from the Audio to Score function can vary given the material Logic is presented to work with. If the source is a clean "monophonic" performance, then the output may well be useable, whereas a polyphonic source (with multiple notes, or several instruments, playing at the same time) with a large amount of background noise will stand little chance of being "transcribed" correctly.

Once edited, it's highly likely that you can use this feature to a create usable "doubling" effect with virtual instruments, or even use it as an alternative way of transcribing played drum parts into MIDI triggers (other than the Drum Replacement/Doubling feature).

Figure 5.49
The Sample Editor's Functions menu contains a number of tools to modify the source data – from the ability to change the gain, through to phase inversion.

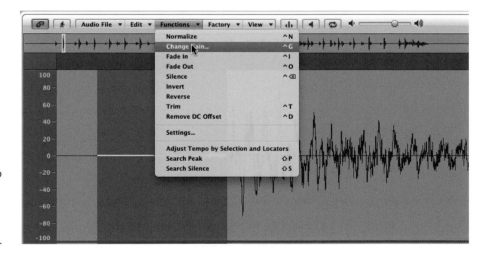

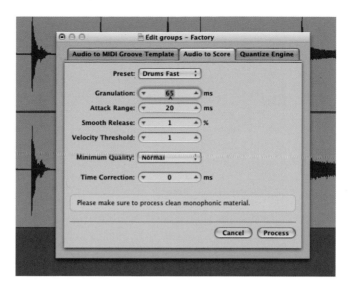

Figure 5.50
Given a clean enough source file, the Audio Score function can perform a passable attempt at transcribing an audio recording as MIDI data.

Time and Pitch Machine

Although we've seen a large amount of "tempo elasticity" courtesy of Flex Time and Apple Loops, it's also worth noting some of the "low-level" ways you can manipulate the pitch and timing of audio files using the Sample Editor. Of course, before the introduction of Apple Loops and Flex Time, the Sample Editor's Time and Pitch Machine was the only way a Logic user could manipulate the speed and pitch of an audio file, although there're still a number of applications that make this a valuable tool to have at your disposal.

As you'd expect, the biggest advantage of the Time and Pitch Machine is its relative accuracy – a timestretch, for example, can be as precise as a few samples, or in the example of a pitch transposition, as fine as a few cents. It's also notable as one of the most flexible ways of adjusting both pitch and time, primarily because it can move between so-called free transposition, where the pitch and time are freely and independently adjustable, and "Classic" mode, which links together pitch and tempo/time changes, working in much the same way as a tape machine's varispeed control.

The Time and Pitch Machine can work either with the region as a whole (selecting Edit > Select All from within the Sample Editor), or within a selected time period within the region. For example, this functionality allows you to individually repitch notes within a vocal line without having to separate each note of the performance into a series of individual regions.

Once you've selected the area to be processed and opened up the Time and Pitch Machine (Factory > Time and Pitch machine), you can then adjust the

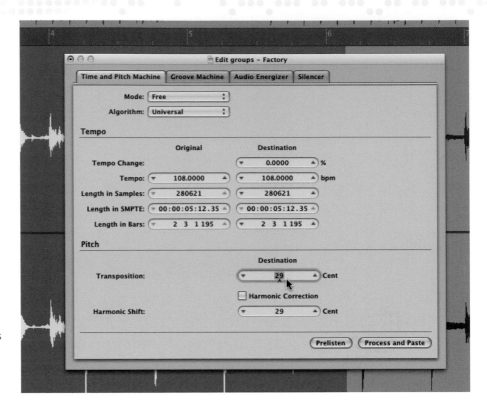

Figure 5.51
The Time and Pitch machine provides one of the most flexible and accurate ways of modifying the tempo, duration, speed, and pitch of an audio file.

relevant parameters in the accompanying dialog box. First of all, you'll need to set the mode – using Free mode to modify the pitch without modifying time (or the other way around), for example, or Classic mode when you want the feature to work much like varispeed (this can sound great on drums, for example). With tempo, you can either specify the change in tempo, a percentage, or the length of the file (either in samples, SMPTE, or Bars), while the amount of pitch change is measured in cents. You might also want to optimize the Algorithm for the material you're working with, especially if the Universal setting isn't delivering effective results.

5.24 Inserting and Deleting Sections of Your Song

One inevitable task of producing a song in Logic is the need to add or remove a segment of time at some point in the project's arrangement – whether you need to remove an unwanted chorus, for example, or add several bars after the third chorus to form the middle eight. Of course, you could perform these tasks using some of the tools and technique we've discussed, or instead, you could

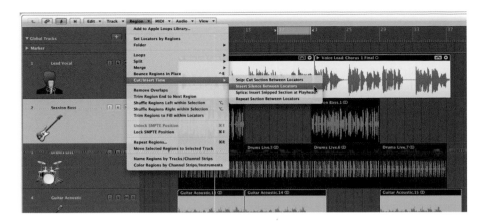

Figure 5.52
Use the Cut/Insert Time feature to extend or shorten your arrangement accordingly.

turn to Logic's specific "Cut/Insert Time" feature to perform this even more quickly and easily!

To take advantage of the Cut Insert Time feature, you'll first need to draw the locators at the top of the Arrange area to indicate the part of the song you want to work with – either the area that needs time "inserted" into it, or the portion of the song you want to remove. Unless you only want the insertion/cut to work with a few selected regions, it's then important that you select all the current regions in the full duration if the project using the Select All key command. Once this is done, use the menu option Region > Cut Insert Time, and choose an appropriate selection (Snip, Insert, Splice, or Repeat).

5.25 Working with Tempo and Loops

Throughout this chapter, we've seen a number of principal ways in which we can deal with the tempo-based material in Logic, especially in relation to the application of Flex Time. However, there are a variety of problems that arise when dealing with tempo-based material, all of which necessitate a unique solution to the problem in hand – whether it's the ability to change your project's tempo based on the tempo of an imported loop, or being able to Beat Map your timeline to a performance that wasn't played to click.

Follow Tempo

Some, but not all audio files will have tempo information embedded within them – this includes Apple Loops files (as we've already seen), but also material you record into Logic, and file formats like ReCycle's REX. In these situations, it's usually as simple as clicking on the Follow Tempo as part of the Region Parameters box so as to have the loop conform to your current project tempo.

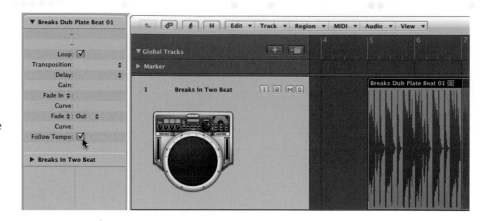

Figure 5.53
If your audio file has tempo information embedded within it, you can use the Inspector' Follow Tempo option to have the region automatically conformed to the project's tempo.

Timestretching a Region to Fit

If your audio file/loop doesn't include tempo information, there're a number of different ways to make it manually conform to your project tempo. First, of course, you can use the Flex Time functionality – engaging a Flex mode and then resizing the region so that it fits precisely to a number of bars. Another option is to use the Time Stretch option under Audio menu. Once you've selected the region in question, pick either Audio > Time Stretch to Nearest Bar, or Audio > Time Stretch to Locators, as appropriate. The option to stretch to Locators is useful if the amount of correction falls outside the "nearest bar." In this case, use the Locators to set the size of the eventual time stretched region.

Working Out the Tempo of a Region/Loop

Another option to manually conform the loop is to use Logic's Detect Tempo feature, which will then allow you to use the Follow Tempo functionality and have an imported loop follow your project tempo. With the region/s selected (in this example, Logic can work with multiple regions to extract a more accurate calculation) select the option Audio > Detect Tempo. In the accompanying dialog box, Logic will provide with a series of "best guesses," which you can select as appropriate, and then check on the Enable "Follow Tempo" or "Flex" to have the resultant calculation used to lock the loop to your project's tempo.

Matching Your Project's Tempo to an Imported Loop

Now let's look at the problem another way around – where you want to change the project's tempo to match that of an imported loop. Once you've imported the region, select it, and then set the locators to match the required bar count – like two or four bars, for example. With this established, select Options > Tempo > Adjust Tempo using Region Length and Locators. Logic will now adjust the tempo

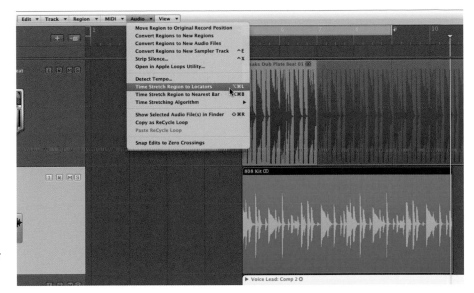

Figure 5.54
If you can't use the Follow Tempo feature, you can turn to manually timestretching the region so that it hits the nearest bar or locator point.

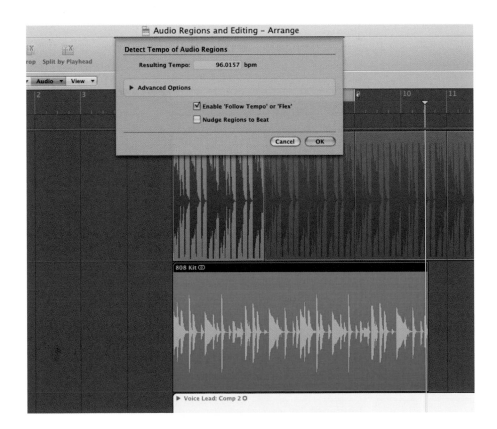

Figure 5.55
Use the menu options Audio > Detect Tempo so that Logic can calculate a "best guess" for the region's BPM. This can then be "flexed" accordingly.

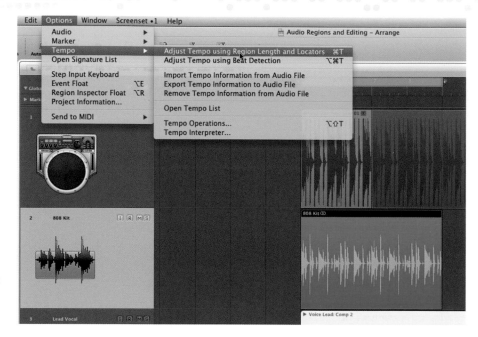

Figure 5.56
As well as getting a loop to play in time with your project, it's also possible to adjust the tempo of your project to conform to the length of your imported region.

so that it precisely fits the duration of the region, and assuming you've set the right bar count, everything should now play in time with the imported loop!

Beat Mapping

As well as working with loops, it's also likely that you'll need to deal with performances that have little or no relation to a rigid metronome or click. In this example, you might want to "beat map" your project's timeline so that it follows the tempo fluctuations of the original performance, theoretically allowing you to synchronize additional loops, or even quantize MIDI parts, to the original recording.

Logic's Beat Mapping feature can be found as part of the Global Tracks feature (which is covered in more detail in Section 10.3). You can view the Global Track by selecting View > Global Tracks. By default, though, the Beat Mapping track-lane isn't displayed, so you'll select View > Configure Global Tracks, and then ensure that the Beat Mapping option is enable.

As you'd expect with Logic, there's a variety of ways you can use the Beat Mapping feature, although in this example, we're going to explore the most common application – that of relocking a prerecorded performance and creating a new tempo grid that follows the fluctuations in the original track.

To get the process started, it helps if you find the average tempo of the performance, as well as editing the first "hit point" so that it falls at the start of

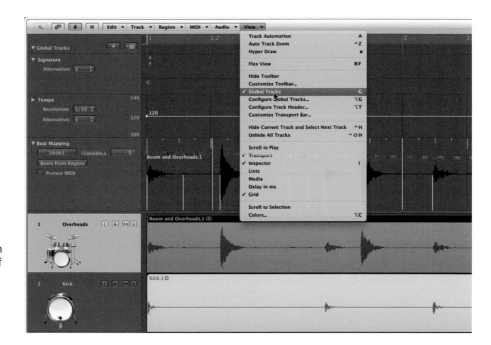

Figure 5.57
Logic's Beat Mapping function is found as part of the Global Tracks feature. To view the Global Track, select View > Global Tracks.

your Logic project. Arguably, the best way of finding the average tempo is to use the BPM Counter plug-in, which can be found under the Metering plug-ins folder. Instantiate the BPM Counter plug-in either across your stereo bus, or across one of the rhythm tracks in question. Once you've established the tempo, use the transport to set this is as your project's global tempo.

Now we can turn our attention to the Beat Mapping process. Expand the Beat Mapping track and highlight the track you feel is most appropriate to work with – which will probably be something like the drum overheads. In the Beat Mapping track, you should now see the waveform of the region you've selected, and, if Logic had already created transient marker information, a series of white transient marker points. If Logic hasn't got any transient marker information, press the Detect button as part of the Beat Mapping tracklane.

Zoom in on the first few bars of the performance – assuming our basic tempo is correct, most (but not all) of the hits should fall on the beat. However, some of the beats will be fractionally off, possibly leading to subsequent beats also falling "off the grid." Using the Beat Mapping feature though, we can realign our tempo grid to that it aligns with the hit points that we feel are important (usually on the bar, for example, or the one of the principle off-beats).

To align the beat, simply click on the beat you want to align to (on the Beat Mapping timeline) and then align this to the accompanying transient point on

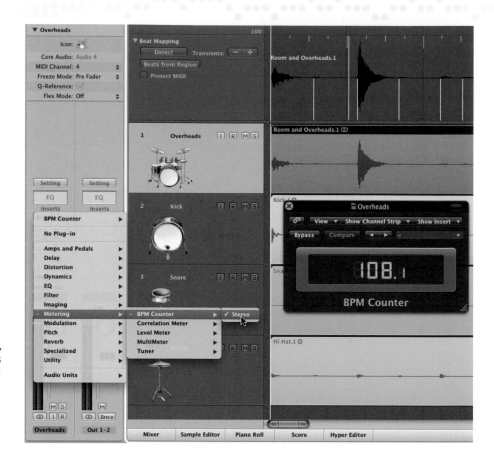

Figure 5.58 To start the Beat Mapping process, align the region/s to the start of the project and find the "average" tempo using the BPM Counter plug-in.

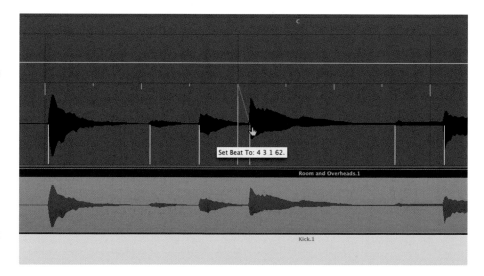

Figure 5.59 Step through the recording and adjust the tempo track to match the transients. First, click on the required beat in the timeline, and second, drag the yellow "beat mapping" line to meet the transient marker in the performance.

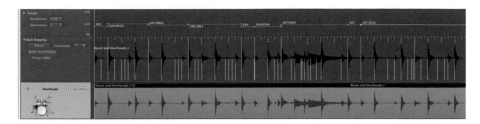

Figure 5.60 Here's a finished example of Beat Mapping in action, with the associated tempo changes to keep Logic in time with the performance.

the Beat Mapping display. As you're aligning the beat and the transient, note how Logic draws a thin yellow line to indicate the alignment process that is taking place. Also, the Tempo track (which is part of the Global Track) should have additional tempo points inserted into it, indicating how Logic is varying its initial tempo to stay in time with the drum performance.

5.26 Managing Your Audio Files and Regions

In these last few sections of our audio editing chapter, we're going to review the techniques that relate to the "management" of you editing workflow – whether it's techniques to organize region data and tracks on-screen so as to make your editing process clearer and easier to negotiate, or techniques to manage the "raw" audio data that is included within your project.

The Audio Bin Tab versus the Audio Bin Window

As we've already seen in Chapter 4, the Audio Bin – found on the left-hand side of our Arrange window – provides a useful overview of the audio files included within our project, alongside basic information about the files in question: such as the sample rate, bit depth, mono/stereo information, and the total size. Clicking on any of the accompanying arrows next to the file opens up the list of regions associated with audio file in question.

However, in addition to the Audio Bin, Logic also includes a dedicated Audio Bin Window (Window > Audio Bin), which allows you to open up the bin separate from the main Arrangement window. Although many of the functions are shared between the two types of Audio Bin, it's worth noting that the dedicated window includes several additional features that make it the ideal choice for more extensive file-based activates – including the ability to display full waveform views for each file/region, as well as an alternative means of setting a region's anchor point (other than using the Sample Editor).

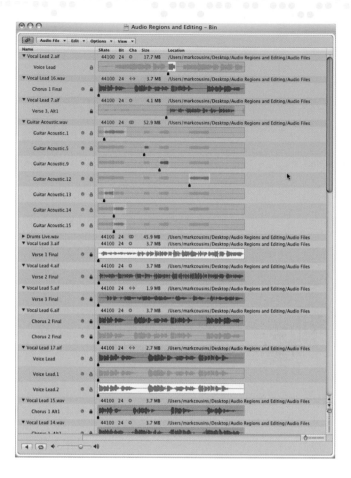

Figure 5.61 The dedicated Audio Bin window provides an expanded set of region and audio file management options, as well as an improved graphical overview that includes waveform data.

File Management

Using a combination of the Audio Bin's Edit menu and the Audio File menu, you can perform a variety of file management activities – whether it's something as simple as deleting an audio file from your hard drive, or moving a batch of audio files to a new location.

One of the most useful features of the Edit menu is the option to Select Unused files. This is a great way of giving your project a spring cleaning – selecting regions and audio files that are unused in the arrangement, which can then be potentially removed from your session using the Delete key. Of course, it goes without saying that you need to approach this function with some care, although its ability to tidy up a particularly messy collection of audio files and regions can be welcome, when used correctly.

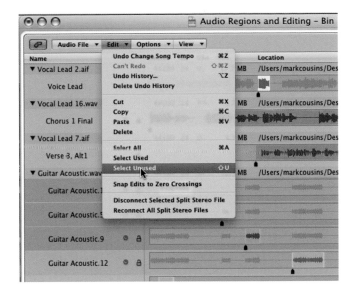

Figure 5.62 Use the Select Unused option to give your Audio Bin a spring cleaning. Once selected, the unused regions and files can be deleted as appropriate.

Using the option Audio File > Move File(s) is a useful way of managing particularly large audio files folders. You can choose to use "nested" folders in a way you feel is most appropriate to a project – this could involve pulling all the vocal files into one folder, for example, or grouping all the files associated with a "take" of a song (assuming you've performed a simultaneous multitrack recording across multiple inputs). The location of audio files also comes in useful with some of the viewing options that we'll see later.

Copy/Converting Files

One important task that the Audio Bin gets used for is the task of converting audio files between different file formats – like AIFF, WAV, or MP3 – as well as moving between different sample frequencies and bit rates. A good example of file conversion in action would be in the task converting an original 24-bit uncompressed mix into MP3 format for use on the Internet. When you choose the conversion option (Audio File > Copy/Convert Files), Logic will present a variety file conversion settings, as well as the option to re-reference the region in your project to the newly created files.

Viewing the Bin and Grouping Files

Using the View menu, you can configure how the Audio Bin displays its list of audio files and regions. One of the most useful features of this menu, though, is the ability to create a Group, allowing you to organize related regions/files

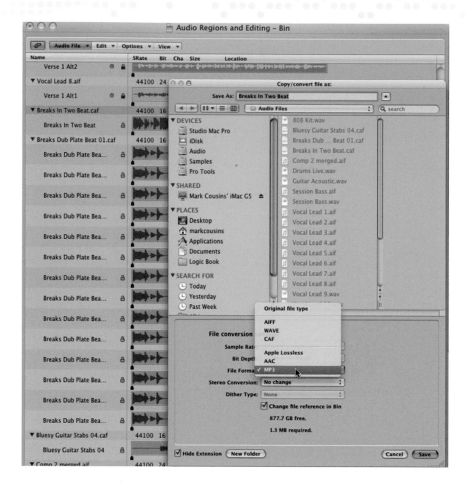

Figure 5.63 The Audio Bin allows you to convert audio files between different bit rates, sample frequencies, and file formats, including AIFF, WAVE, and MP3.

Figure 5.64 Use the Audio Bin's Group Files feature to better organize how it displays your audio files and regions.

so that they appear together in the Audio Bin. You can create a Group in a number of ways – either by selecting the files manually and then choosing View > Create Group, or by using one of the automatic group options. For example, you could group files by their location on your hard drive (assuming you've used the "nested" folder previously described), their file attributes, or the current selection in the Arrange area.

5.27 Using Folders and Hiding Tracks

In modern music production, it's common to have a number of "related" tracks and regions – maybe the close and ambient microphones placed around a sing drum kit recording, for example, or the multiple performances that form a "chorus" of backing vocals. As these tracks and regions amass, it can get difficult to manage the amount of material on screen, especially if you're working on a smaller-sized monitor.

Logic's Folder feature, therefore, is an elegant solution to this problem – allowing you to group together related regions and tracks and pack them inside a folder. With the regions packed inside a folder, you can enjoy a more simplified arrangement area, with the mass of regions now represented as a single region/folder. At any point in the production, you can "peer inside" and see the contents of the folder – editing its contents accordingly, but only seeing the regions and tracks that relate to that folder. Move back out of the folder, and you'll find yourself back on the "root level" of your arrangement.

Of course, the Folder feature bears some similarities to Take Folders used as part of the Quick Swipe Comping feature. Despite these apparent similarities, though, it should be noted that a Take Folder is principally used for the task of comping, whereas a Folder can be used to organize a number of discrete and separate regions working independently of one another. In short, use Take Folders for multiple takes, and Folders for groups of related instruments/regions.

Packing a Folder

To pack a folder, first select the regions you want to be include in the folder, and then choose Region > Folder > Pack Folder. Logic will now create a new folder (which looks and behaves much the same as a region) as well as a folder track to play this back. Both the folder and track can be named as you see fit.

To clear some space in the arrangement, you might also want to use the menu option Track > Delete Unused. This will remove the empty tracks from the Arrange area (left when the regions where placed in the folder), although of course, they're still active "inside" the folder.

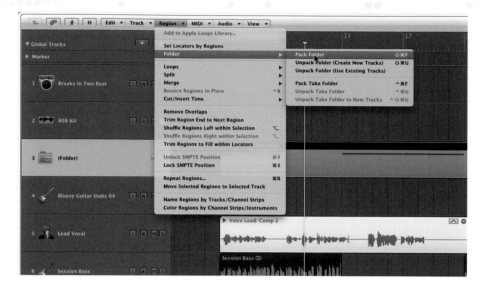

Figure 5.65
To organize complicated arrangements with multiple regions and tracks, use the menu option Region > Folder > Pack Folder.

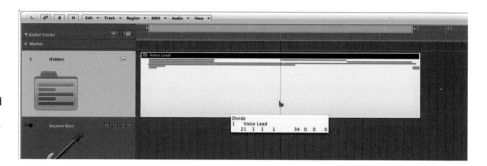

Figure 5.66
A folder works much the same as a region, allowing it to be resized, copied, and muted in much the same way as an ordinary region in the Arrange area.

Despite containing a number of regions, a folder can be edited and copied in much the same way as an ordinary region – resizing it, for example, to make it shorter, or duplicating the folder so that appears more than once in the song. As such, a folder is a very efficient way of handling multiple regions, especially where lots of edits have been made and there's a risk of losing small region fragments as you move large selections of regions around your arrangement.

Viewing a Folder's Contents

You can view the contents of a folder by double-clicking on it. Logic will then present an arrangement solely of the regions (and associated tracks) that are contained within it. This will also be reflected in the Mixer, which will also display a slimed-down set of tracks. To move back up to the "root level" of the arrangement, simply click on the back arrow in the top left-hand corner of the Arrange area.

Figure 5.67
Press the back
arrow in the top
left-hand corner of
the Arrange area
to move out of the
folder and return
back up to the
root level of the
Arrange area.

Of course, at any point in the production process, you can choose to unpack the contents of the folder, using either the menu option Region > Folder > Unpack Folder (Create New Tracks), or Region > Folder > Unpack Folder (Use Existing Tracks) accordingly.

Hide Tracks

As an alternative to packing folders, you can also choose to, temporarily, hide a number of tracks from the current view. The Global Hide View Button can be found in the top left-hand corner of the Arrange area, and when depressed a series of buttons also inscribed "H" appear on each track beside the record-enable button. Simply choose the tracks you wish to hide, and then click on the "H" button at the top of the Arrange area. The tracks you've selected to Hide will now be out of view (and the Global Hide View button illuminated orange to indicate this), allowing you to concentrate on a simplified arrangement. When you need the tracks returned back into view, simply press the Global Hide View button once more.

In This Chapter

6.1 Introduction 145

6.2 MIDI Concepts 145

6.3 Creating Instrument Tracks 146

6.4 Instantiating Virtual Instruments and the Library Feature 147

6.5 Working with External MIDI Instruments 152

6.6 Making a MIDI Recording 158

6.7 Editing and Arranging MIDI Regions 161

6.8 Region Parameters: Quantizing and Beyond 162

6.9 The MIDI Thru Function 166

6.10 Advanced Quantization Options 166

6.11 Normalizing Sequence Parameters 169

6.12 MIDI Editing in Logic 169

6.13 The Piano Roll 170

6.14 Typical Editing Scenarios in the Piano Roll 172

6.15 Quantizing Inside the Piano Roll Editor 178

6.16 Working with Controller Data Using Hyper Draw 180

6.17 Hyper Draw in the Arrange Area 182

6.18 Going Further: The Piano Roll's Edit and Functions Menus 183

6.19 Intelligent Selection: The Edit Menu 183

6.20 Functions: Quick-and-Easy Note Modifications 186

6.21 Step-Time Sequencing 188

6.22 The Hyper Editor 190

6.23 Score Editor 193

6.24 Event List 193

Knowledgebases

EXS24 Virtual Memory and Disk Streaming 176

Using Multioutput Instruments 188

Separating MIDI Events 191

EXS24 Data Management 196

Plug-In Focus

EVP88 149

EVB3 154

EVD6 160

ES E, ES P, and ES M 165

EXS24 Main Interface 171

MIDI Sequencing and Instrument Plug-Ins

6.1 Introduction

Besides being a tremendously competent and versatile audio sequencer, Logic has been famed for its abilities at recording and manipulating MIDI data, albeit in the virtual domain – using Logic's own virtual instruments and other third-party plug-ins – or in the now-dwindling world of hardware synthesizers and samplers. Working with MIDI in Logic offers some of the most creative and flexible musical production tools available, offering complete control of every last musical detail in your track: from the precise time and placement of notes to the exact timbre and sound parts are played with.

In this chapter, we're going to explore some of the key techniques of working with MIDI, from basic region manipulation on the arrange page to more in-depth editing techniques using Logic's various MIDI editors: the Piano Roll, Hyper Editing, and the Event List. We'll also explore some of Logic's key virtual instruments – including the EXS24, the CPU-light plug-ins ES1, ES E, ES M, and ES P, alongside the EVP88, EVB3, and EVD6 vintage instruments – all tools that you'll undoubtedly use on a daily basis. If you're particularly interested in the creative application of MIDI, and the techniques of sound design, be sure to read Chapter 7, which follows on from many of the concepts dealt with here.

6.2 MIDI Concepts

MIDI recording in Logic involves working with either virtual instruments embedded within the application itself (like the ES2 synthesizer or the EVP88, a modeled electric piano) or MIDI equipment, like hardware samplers and synthesizers, externally connected to Logic via an appropriate MIDI interface. Unlike audio recordings, a MIDI recording makes an important detachment between musical data (in other words, the raw note information that forms a musical performance) and patch data (the information required to generate the sounds). Fundamentally, what this means is that within the MIDI sequencer, we

DOI: 10.1016/B978-0-240-52193-0.00006-X

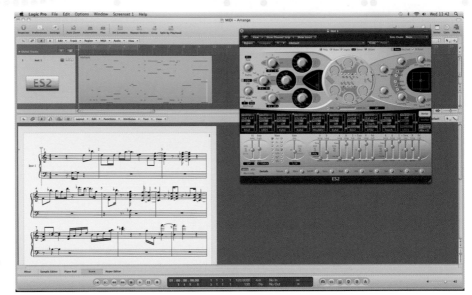

Figure 6.1 Unlike an audio region, a MIDI region only contains the note data, with a virtual instrument (or MIDI synthesizer) used to control the precise properties of the sound.

have complete and independent control over each of these elements – we can change the sound or the performance in subtle or radical ways at any point in the production process.

6.3 Creating Instrument Tracks

Creating a suitable MIDI track is done via the New Tracks command (as we saw with the creation of an audio track), although in this example, we have the choice between a Software Instrument (which would be used to access any of Logic's own virtual instruments or third-party audio units like Native Instruments' Absynth or Kontakt) and an External MIDI track, in the example of accessing legacy MIDI hardware.

The options presented for the new track types vary to that of an audio track, with a few less parameters to choose from. As with audio tracks, you can create multiple instances, which is handy if you know you're going to require a number of different virtual instruments in your session (16, for example, should provide an adequate starting point). If you're using third-party software instruments like Kontakt 4 that feature multitimbral operation (in other words, you can load multiple sounds into the instrument), you might want to select the multitimbral option. This will create 16 different tracks, all of which are assigned to the same instrument and will allow you to address the 16 different channels, or instrument slots, within it.

Figure 6.2 Use
the New Tracks
command
to create a
new Software
Instrument
track – for either
Logic's own virtual
instruments or
third-party audio
unit plug-ins like
Kontakt 4.

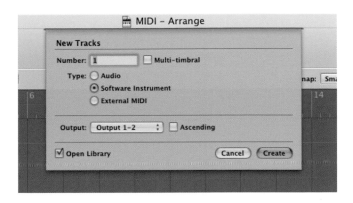

6.4 Instantiating Virtual Instruments and the Library Feature

Assigning a virtual instrument to a corresponding Software Instrument track is done in one of two ways – either selecting from a wide range of so-called Channel Strip Settings or selecting a specific software instrument that interests you.

Let's start by taking a look at loading in case of a specific software instrument – albeit one from Logic's integral instrument collection or a third-party audio unit. Possibly, the easiest way of doing this is via the Channel's Inspector using the small mixer strip that appears in the bottom left-hand corner or the Arrange window. By clicking just below the I/O label (in what should currently be an empty grayed-out box), you'll open the list of Logic's integral software instruments, and at the bottom of the list, an option to browse AU instruments, which should have all your third-party audio unit instruments organized by manufacturer.

Once the instrument has been instantiated, its interface will be presented to you on the screen. On the top of the interface, you should find a drop-down list of the various presets associated with the instrument, as well as two arrows to sequentially move backward and forward through the list.

Browsing Instrument Settings in the Library

As we've seen in previous chapter, the Media area can be used to browse a list of audio files in our session, the Apple Loops collection, and indeed, your computer's hard drive. Another extension of this is the Library feature, which offers an alternative, list-based way of browsing and navigating an instrument's settings. To access the Library, open the Media area (B) and click on the Library tab. You should now see a list of your presets settings, possibly organized into folders, through which you can navigate as you see fit.

Figure 6.3 Parameters and presets can both be directly controlled from the instrument plug-in's interface.

Figure 6.4 The Library provides a more elegant and seamless way of stepping through an instrument's settings.

Plug-In Focus 1 ▼

EVP88

The EVP88 was one of the first virtual instruments developed by Logic, modeled on the classic electric piano sounds – like the Fender Rhodes and Wurlitzer – used throughout the 1960s and 1970s. Of all the vintage instrument plug-ins included in Logic, the interface is one of the easiest to comprehend, with a relatively simple set of parameters to work with. The most important of these is the Model parameter, which selects a basic model from which to work. Models range from the classic Rhodes designs, like the Suitcase Mk1 and Suitcase V2, to Wurlitzers (great on rock tracks) and a Hohner Electra.

The parameters found toward the bottom of the interface replicate the more traditional controls and features found at an electric piano player's hands. EQ and Drive, for example, govern the basic tone of the patch – keep these relatively flat if you want a clean DI'd sound or ramp them up if you need to produce the more colored tone, like the original Rhodes fed through a fender twin amp. The Tremolo control is also another feature integral to the sound of the original pianos, with controls to change the rate and intensity of the modulation. Try experimenting with the stereo phase – the 180° setting produces a pleasing stereo autopan, while setting the modulation

Figure 6.5 The EVP88 remains one of the silkiest electric piano emulations available, with a surprisingly low CPU drain.

(Continued)

progressively nearer 0° produces an increasingly mono Tremolo. The Chorus and Phaser are also important Rhodes-orientated effects, used to great effect in the late 1970s on tracks like 10cc's "I'm Not In Love."

Experimenting with the model parameters allows you to twist certain "mechanical" properties of the original instruments – variations that might have occurred with the condition of the instrument, for example, or in the way it was manufactured. For example, increasing the Bell volume makes the sound closer to the bright "Dyno Rhodes" modifications performed on Fender Rhodes pianos in the early 1980s.

Using Channel Strip Settings with Instruments

As an alternative to instantiating a single instrument, you can use Logic's Channel Strip Settings feature. A Channel Strip Setting offers the complete range of settings for the current selected Instrument tracks – instantiating an appropriate instrument and loading samples (where appropriate), alongside the complete set of plug-in effects (reverbs, equalization, compression, and so on) that are associated with that instrument.

Arguably, what you get with the Channel Strip Settings feature is an output that's closer to a finished sound as it might appear in the mix, although it's worth remembering that these particular settings might not be entirely appropriate to your eventual balance of sounds. Another potential downside of the extra plug-ins is the increased drain on your CPU – for example, it might be more efficient

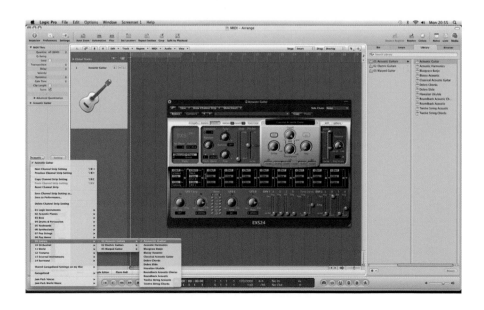

Figure 6.6
Channel Strip Settings provide a complete combination of instrument and associated effects. This is a useful solution for mix-ready sounds, but the effects might begin to clutter the mix.

for several instruments to share a reverb plug-in rather than each instrument having its own reverb setting. Of course, there's always the option of disabling plug-ins (by Alt-clicking on them or activating their bypass control), or even removing them entirely, should you change your mind at a later point.

The Channel Strip Settings can be accessed via the small gray Setting box at the top of a channel strip. Note that if you've got the Library open, clicking on the same Setting box will also open up the Channel Strip Settings in the Library browser.

Of course, the Channel Strip Settings can also be used if you want to save your own Instrument settings with their associated effects – maybe to move a patch between different project files, for example.

Changing the Instrument Parameters

Instrument Parameters are a global set of parameters associated with a particular instrument. The Instrument Parameters can be found via the Inspector, with the current selected track having its parameters displayed toward the middle of the Inspector – remember to open the box (using the small arrow) if it is minimized. Although the Instrument Parameters include some useful features, most users tend to leave these settings in their default positions, unless they're working with external MIDI sources (see Section 6.5 for more information on this).

Changing an Instruments Icon

Icons allow you to easily distinguish between different tracks in your Logic project. By default, any of Logic's instruments will establish a corresponding icon to distinguish between the different models (EXS, ES2, and so on), although you might want to use one of Logic's other icons to define a track's musical properties (strings, bass, guitar, drums, and so on). To change the icon, simply click on the icon setting in the Instrument Parameters box.

Transposition and Velocity

Use the transposition and velocity parameters to quickly change the way an instrument responds to the performance data it's presented with. For example, if you know you want the instrument to take a prominent, aggressive position in the mix, it might be appropriate to use an additive value (like 120) for the velocity setting. All performance data sent to this instrument – whether played live from MIDI keyboard or from any track on the Arrange window assigned to this instrument – will be modified accordingly. If you want more control as to when these modifications occur, consider applying them as MIDI edit – either by modifying region parameters, or by using one of the MIDI editor windows, both of which we'll explore in more detail later on in this chapter.

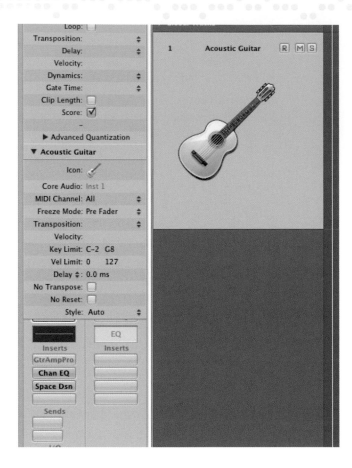

Figure 6.7 The Instrument Parameters box allows you to control a number of key parameters in relation to the instrument instantiated into a particular track. On the whole, this can be left in its default state.

Key Limit

This feature is useful where you're orchestrating for acoustic instruments with a known limit in their playable range. For example, a first violin could have its lower range set to G2 to remind you not to play or record any MIDI data below this point. Although some of the sample instruments included with the EXS24, or as third-party data, might account for the natural instrument's playable range, not all do. A little forethought here, therefore, could save hours of extra editing and arrangement work later on.

6.5 Working with External MIDI Instruments

First off, if you haven't got any external MIDI equipment, you can skip this section and carry onto the next! However, if you're lucky enough to have some external MIDI sound sources, like (proper) vintage synthesizers and samplers,

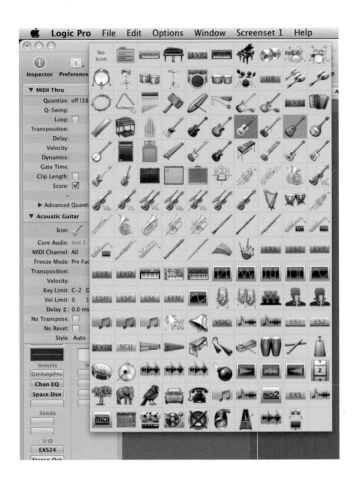

Figure 6.8 Using icons will allow you to better distinguish between different tracks and their various musical roles.

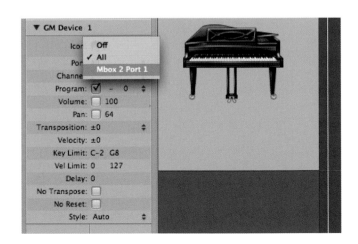

Figure 6.9 Change an External MIDI Instrument's Port setting (as part of the Instrument Parameters) to assign it to a physical connection on your MIDI/audio interface.

then you'll need to spend a few minutes adjusting a few settings to get things working correctly. It could also be the case that you're also using the MIDI protocol to communicate with other computers in your studio, even using a network connection to connect to something like GigaStudio, in which case you'll also need to follow these basic configuration steps.

Because we're working with an external MIDI device, we'll need to ensure that our track is being directed to the right physical MIDI output on our corresponding MIDI or audio interface. Start by selecting the External MIDI track in question, opening the Inspector, and looking at the Instrument Parameters. You'll need to establish a Port setting to correspond to the port your MIDI device is plugged into, as well as setting a corresponding MIDI channel. If the instrument is multitimbral – in other words, a device that allows a different sound to be placed on each of the 16 available MIDI channels – you might want to run multiple External MIDI tracks, each set to a different MIDI channel. To do this, go to the Track menu and select New with Next MIDI channel (or Apple + Alt + M). You can also Ctrl-click at any point on the track name to reassign the object to another MIDI channel accordingly.

Plug-In Focus 2 ▽

EVB3

The EVB3 is Logic's own virtual recreation of the classic Hammond organ. The original instrument used a unique tonewheel sound generation system, controlled by a series of drawbars, to create a surprisingly rich and varied set of organ sounds. The package was then made complete by the addition of a rotating speaker cabinet – otherwise know as the Leslie – providing a distinctive swirling movement to any sound that passed through it.

Like the original B3, the EVB3 supports two manuals (or keyboards) and a set of bass pedals, each with its own set of drawbars. To access these, you'll need to set the MIDI channel in the instrument's parameter box to All – this way the EVB3 will be able to receive MIDI data on all channels, with Channel 1 being directed to the upper manual, Channel 2 to the lower manual, and Channel 3 to the bass pedals. Alternatively, change the Keyboard Mode from Split – you should now find the various manuals accessible from different octaves of your controller keyboard.

Shaping the tone of the EVB3 is achieved by sliding the relative position of each manual's drawbars. Essentially, the drawbars represent a different harmonic from the harmonic series – try using a simple combination of just a few of the lower drawbars for a mellow, pure tone, or "pull out all the stops" for a fuller timbre that is rich in harmonics. The colors also provide some indication to their respective roles: the white drawbars being even harmonics, black drawbars containing the odd harmonics, and the brown drawbars representing the sub (an octave below the fundamental) and third harmonic.

Switching between the different Leslie settings (Chorale, Brake, and Tremolo) forms a distinctive part of the movement and progression in a Hammond performance. The Chorale creates a slow vibrato, the Tremolo a fast-moving vibrato, whereas the Brake stops the vibrato effect. In addition to being able to switch the settings from the interface (these could, for example, be recorded into MIDI), you can also use the mod wheel to move through the three settings.

Figure 6.10 The EVB3 is Logic's take on the classic Hammond organ, with two manuals and a set of bass pedals, and of course, the authentic drawbar sound creation system.

How Do I Monitor MIDI Sound Sources?

Although you may have routed MIDI out of Logic to the device in question, there's still no guarantee that you'll be able to monitor it correctly. Unlike Audio Instruments, of course, an external MIDI instrument won't have a corresponding mixer channel to monitor its output (although you do have a fader to remotely control the level and pan settings via MIDI). One solution is to run your main outputs from Logic into a physical mixer (Channels 1 and 2), with your synthesizer, sampler, or drum machine connected into any remaining spare channel. With this setup, however, you won't be able to use any of Logic's own effects on the synthesizer's output, or indeed, use the bounce to disk mixdown functionality to create a complete version of the mix.

A more appropriate solution is to plug the outputs of your MIDI hardware sound sources into spare inputs on your audio interface. Next, create a number of spare audio tracks (one for each device connected to your audio interface), with each of the audio tracks set to Input Monitor. Return back to your External

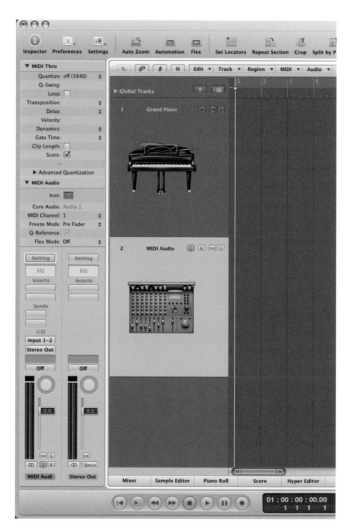

Figure 6.11 To monitor a MIDI synthesizer, you'll need to connect it to your audio interface's audio inputs, create a corresponding audio track, and place this into Input Monitor mode. This will also eventually allow you "print" the synthesizer's output as an audio file.

MIDI track, press a few keys on your controller keyboard, and you should now hear the sound source being routed back into Logic's main mix. Note that you might need to adjust the input gain on your audio interface to account for the levels being sent out from your sound source, and that the monitoring will, of course, be subject to any latency present in your system.

The significant advantage of your MIDI sound module being routed in this way is that you have complete control over how it appears in the Logic mix, even with the possibility of adding plug-ins, like equalizers, compression, and reverb. Of course, if you're completely happy with the performance and sound, you can quickly flick from Input Monitor to Record and have the part permanently stored as an audio file, which might offer a better long-term solution, as well as being a convenient means of negating any latency issues.

What If I've Got More Than One External MIDI Device?

Adding further MIDI devices beyond the basic GM Device included in the basic Empty Project file goes beyond the scope of the main Arrange window interface. Instead, you'll need to create new instrument objects as part of the environment, which is something we'll cover in more detail in Chapter 11. If you have a particularly complicated MIDI setup, you might want to consider building your own Template file so that all your preexisting MIDI devices are automatically set up with each new project. Again, this is something we'll specifically look at in Chapter 11.

My MIDI Tracks Are Playing Back Slightly Late, How Do I Rectify This?

As we mentioned before, it is feasible that the monitoring of External MIDI tracks via your MIDI hardware could be causing some minor latency issues, with the MIDI tracks appearing to play slightly behind the beat. To rectify this, you might want to consider applying a negative delay via the Instrument Parameters so as to send the MIDI data out slightly ahead of time and therefore negate any of the latency involved. Finding the right settings (which will probably be in the region of about – 20 or so, depending on the latency in your system) could be as simple as adjusting the delay parameter by ear. Alternatively record in a bar or so of MIDI information as an audio file, and then adjust the instrument's negative delay until the "live" MIDI data and recorded MIDI data playback exactly in sync.

One important factor to note here is that once a MIDI track has been recorded as audio, the monitoring latency will be removed. It is important, therefore, that if you intend to record an external MIDI part into Logic, you ensure that the delay parameter has been reset to its zero point.

To throw yet another wrench in the works, it should also be pointed out that Logic's plug-in delay compensation (found under Preferences > Audio > General),

Figure 6.12
Adjust an Instrument's delay settings to account for any monitoring latency.

where used with latency-including DSP-acceleration systems like Universal Audio's UAD-1 or TC Electronics' PowerCore, can also create additional MIDI playback latencies that will vary given the amount of plug-ins, and where they are placed in Logic's mixer. The best advice here is to avoid applying these types of plug-ins until the mix or use the low-latency monitoring mode to temporarily bypass the offending plug-ins while you are monitoring External MIDI instruments.

6.6 Making a MIDI Recording

Making a MIDI recording follows many of the same procedures explored in Chapter 4, although you might want to note a few dissimilarities.

Where's the Data?

The MIDI data created with a recording stays within the region, rather than referencing a separate audio file, as is the case with an audio region. MIDI data takes up an incredibly small amount of memory, with a whole project's MIDI resources utilizing less than about 1 MB of RAM in total.

Recording in Cycle Mode and Overdubs

Unlike an audio recording, which will create a Take Folder for parts recorded in Cycle mode, a MIDI recording will simply layer addition passes into the same region. The assumption with Cycle mode is that you use the repeating playback to record each part a layer at a time (like a kick, followed by the snare), without stopping the transport at any point. If you want to record different takes of a keyboard solo, for example, consider creating a new track assigned to the same instrument (Track > New with Same Channel Strip/Instrument), with the previous take muted.

Figure 6.13
Multiple MIDI recordings will simply create "layered" takes, which can get confusing on a single MIDI track. Consider creating duplicate Instrument tracks to better distinguish between your various takes.

Figure 6.14 In this example, a duplicate track has been used to record the second take of a MIDI performance, rather than layering it onto the original MIDI region.

Recording directly over the first take (as you would with an audio overdub) will simply create regions placed on top of one another, all of which will play at the same time. Multiple stacked MIDI regions – unlike Take Folders – can get difficult to manage, so try to keep on top of any layered recordings using the editing and arrangement techniques described in the following parts of this chapter.

Plug-In Focus 3 ▼

EVD6

The EVD6 is Logic's virtual recreation of Hohner's "funkalicious" Clavinet. The original instrument worked like an electrified harpsichord, with a number of reeds plucking a series of strings, with the resultant vibrations being picked up by a collection of guitar-like pickups mounted in the unit.

As with the EVP88, the EVD6 divides its interface up into areas with more conventional controls – as found on the original instrument – as well a detailed set of modeling parameters to adjust every last detail of the EVD6's output. The conventional controls focus on the rocker switches toward the bottom of the interface. These move between basic, preset filter positions (brilliant, treble, medium, and soft) and different pickup combinations. For example, by alternating the various filter rockers, you can move between a soft, almost muted tone, to a full and raspy clavinet pluck. The pickup switches produce more subtle transformation in the sound and are worth experimenting with to extract just the right flavor from the instrument.

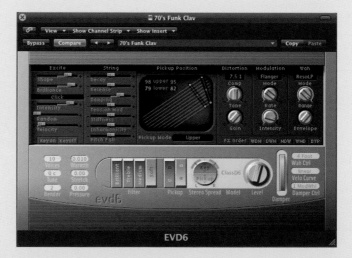

Figure 6.15 Logic's EVD6 is a faithful recreation of Hohner's "funkalicious" Clavinet – complete with the original controls and the number of modeling options.

Along the top of the interface are the various modeling parameters – including some interesting options to physically move the pickup positions inside the virtual Clavinet – and the various additional FX options. In keeping with the approach of many keyboard players in the 1960s and 1970s, the EVD6 includes a Distortion section, modulation effects (Phaser, Flanger and Chorus), and the infamous wah-wah. Although the wah-wah can be set to track the sound's envelope (for an autowah effect), the best solution is to map it to something like the modulation wheel, or a spare MIDI controller, to keep it as hands-on as possible.

6.7 Editing and Arranging MIDI Regions

Again, the process of editing and arranging MIDI regions on the Arrange window follows many of the same conventions as audio editing, allowing you to sketch out and develop your music arrangement over time. As we saw in the audio editing chapter, regions can be resized, cut, moved, and duplicated using the same tools and processes involved in audio editing. However, as with the process of recording, it's worth pointing out a few differences with respect to working with MIDI regions.

Merging MIDI Regions

Using the glue tool allows you to merge a number of MIDI regions into a single contiguous region, although unlike audio data, this doesn't result in the creation of a newly rendered audio file. Merging MIDI regions is a great way of combining any stacked MIDI regions (where regions have been stacked on top of one another), allowing you to see all the MIDI data in a unified way.

Working Without Fades and Crossfades

As you'd expect, a MIDI region doesn't require any crossfades to be placed at edit points; therefore, it is impossible to apply fade-based processing to a MIDI region. To achieve the same volume contouring, you'll need to turn to either Automation (for more information see Section 8.10 in Chapter 8) or the Hyper Draw feature (covered in Section 6.17).

Timestretching MIDI Regions

Although it is possible to perform a certainly level of time adjustment with audio regions, it is MIDI regions that offer some of the greatest flexibilities with respect to time-based properties. By Alt-dragging the end point of a MIDI region, for example, you can quickly time compress or expand the data that are

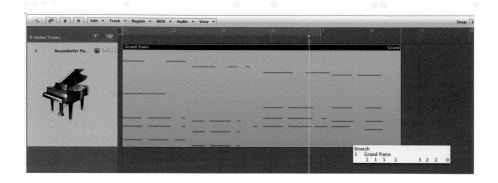

Figure 6.16
Using the Alt key as you resize a MIDI region will allow you to time compress or expand its MIDI data accordingly.

contained within it. This can be an interesting way of exploring double-time or half-time treatments by stretching the region by 200 or 50% of its original size. Of course, as this is simply the playback of MIDI data you're effecting, there is no loss to the sound quality of playback, which would easily become apparent with the same level of stretch applied to an audio file.

It should also go without saying that a MIDI region follows and responds to the project's tempo, without needing to activate the Follow Tempo feature in the Region Parameters box. This also makes it feasible to record a MIDI region at more comfortable playing speed and then increase Logic's tempo to playback the region at the appropriate speed. The only issue to be aware of here, though, is accompanying audio tracks, which won't necessarily follow the same tempo change. In this case, simply monitor from virtual instrument or MIDI tracks accordingly.

6.8 Region Parameters: Quantizing and Beyond

Once you've recorded your first few bars of MIDI data, you can now start to explore some quick-and-easy modifications directly from the Arrange window. One of Logic's key concepts is that a large number of MIDI modifications can be performed directly from the Arrange window, without ever having to go anywhere near a conventional MIDI editor (like the Piano Roll or List editor windows). To perform these modifications, simply select one or more regions and turn your attention to the Region Parameters box as part of the Inspector.

The main parameters accessible from the Region Parameters box include the region's name, quantizing, loop, transposition, velocity, dynamics, gate times, and delay settings. What the Region Parameters allow you to do is to quickly define some important overriding concepts of how the region – or group of regions – is played back. No data is modified – these parameters are simply "playback" criteria. For example, you could select the entirety of your MIDI

Figure 6.17
The Region Parameters – available via the Inspector – allow you to quickly define key properties of your MIDI regions.

composition and adjust the transpose option to find a key that best suits your vocalist's range. Should you decide to return your composition to its original pitch, simply select the regions and reset the transposition accordingly.

Quantize

Arguably, the most important function of the Region Parameters is the Quantize feature, allowing you to quickly tighten the rhythmic qualities of a given performance. Clicking toward the right-hand side of the Region Parameters box allows you to open up a list of the Logic's available quantize options, ranging from various 1/8, 1/12, and 1/16 note settings, to Swing quantizing, and combined 16/24 options. On the whole, most conventional musical applications will adopt a simple 1/16th note quantizing so that any stray notes are pulled to a rigid sixteenth (or semiquaver, to give it its musical name) grid. A Swing quantize, on the other hand, shifts every other sixteenth, producing a distinctive shuffle effect commonly used in Hip-Hop or R n B productions. In Swing quantize, the various letters (16A, 16B, 16C, and so on) denote the relative amount of shuffle, with 16A, for example, delivering a lighter shuffle than 16D.

Q-Swing

The Q-Swing offers a finer control over the swing setting than the 16A, 16B, 16C settings found in the standard drop-down Quantize menu. At 50%, the feel is straight, while settings either way (60% or 40%) imparting different qualities of shuffle. If you're particularly into the groove of a piece of music, then the precise control offered here will be important in creating the right feel between parts.

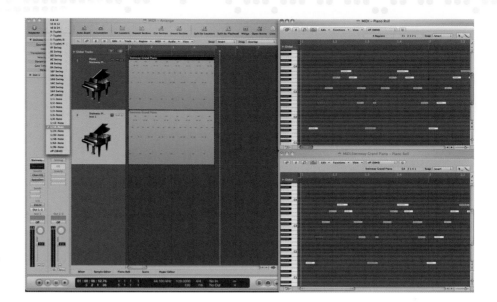

Figure 6.18
Here's a simple MIDI sequence before and after quantizing. Note how the timing of the quantized region (top) conforms exactly to the project's tempo grid.

Loop and Transpose

Loop – as with the same function in an audio region's Region Parameters – can set an indefinite loop of the MIDI region until either the end of the project or another sequence object is encountered. Transpose, of course, deals with the relative pitch of the MIDI sequence, allowing you to transpose a region, or group of regions, up or down by a selected amount of semitones.

Velocity and Dynamics

Velocity and dynamics allow you to control the relative ferocity that a part is played at – setting a number in the velocity parameter, for example, will either add or subtract a value to the current velocity levels contained within the part. The dynamics parameter is even more interesting, in that it works like a MIDI compressor or expander – reducing or expanding the range of velocity levels that occur in the region. Indeed, couple these functions with an expressive virtual instrument, and you can have a surprising impact on how the instrument sits in your mix.

Gate Time

Gate time deals with the duration of notes – changing an elongated series of notes, for example, into a shorter staccato effect. You can achieve some interesting effects here on repetitive sequencer lines, especially where all notes are of fixed duration. Try experimenting with the different gate times and see how this affects the finished output.

Plug-In Focus 4 ▼

ES E, ES P, and ES M

Although it's easy to get carried away with many of Logic's deeper and more complex software instruments, there's still a lot of use to be extracted from some of its simpler synthesizers. One of the best qualities of these instruments is their simplified user interfaces, often with a bias toward specific sound types, like pads in the case of ES E, or leads and basses in the case of ES M.

Arguably, the simplest synthesizer in Logic is the ES E, which is dedicated to simple pads sounds, all from a single oscillator. Try experimenting with some of the key parameters (in other words, the ones with the bigger knobs), exploring the different waves on offer, and tools like cutoff and resonance. The three modes of Chorus (Chorus I, Chorus II, and Ensemble) are vitally important in creating warmth to the pad sound, with each setting offering a slightly different flavor.

ES M is a virtual instrument modeled on the simple synthesis capabilities of early monophonic synthesizers, like the Roland TB-303. This is a great source

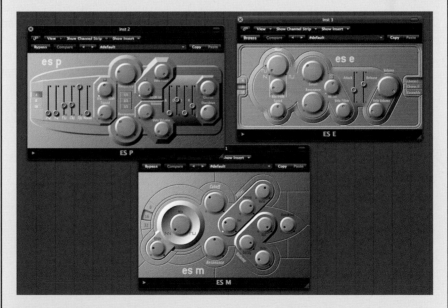

Figure 6.19 Logic's CPU-light collection of synths can be a great place to turn to for some simple keyboard sounds: like 1-oscillator synth basses or sawtooth pads. Note how each synth is dedicated to a certain range of sounds, making the interface uncluttered and easy-to-use.

(Continued)

for "acidic" basslines, full of distortion and resonance, as well a simple lead synthesizer parts. The Glide parameter is an interesting feature – and a quintessential part of the aforementioned TB-303 bass sound – allowing the ES M to gliss from one note to the next whenever the first note is held over.

The ES P offers the most complex set of sound possibilities of the three synths and can create a range of different synth sounds. One of the most distinctive features is the waveform mixer, which allows you to blend a number of different waveforms to create a starting point for the rest of the synth. The first three sliders govern basic waveform types (triangle, sawtooth, and square), while two of the final three sliders add square waves an octave below and two octaves below that of the fundamental pitch. Further sound-shaping features include envelope generators, Chorus, and an overdrive section.

6.9 The MIDI Thru Function

If no region is selected in the Arrange area, you should notice that the Region Parameters box displays MIDI Thru rather than an appropriate region name. As innocuous as the MIDI Thru function may seem at first, it is, in fact, a particularly powerful feature of sequencing in Logic. MIDI Thru's most immediate and noticeable benefit is as "live" modifier of MIDI data coming into Logic. For example, rather than playing in a difficult key, like C# major, you could choose to play in C major and use the transposition parameter to shift everything that you play up half a semitone. Equally, the velocity parameter can be a useful way of increasing or decreasing the relative loudness as you play.

Another important function of the MIDI Thru option should also become evident as you make recordings into Logic. In effect, the MIDI Thru function sets a default Region Parameter setting for everything you record into Logic. Going back to the transposition example, you should note that a part recorded in with MIDI Thru set to 11 transposition will have its Region Parameter set accordingly (this also means you can go back to the data in its original form). This MIDI Thru functionality is particularly powerful with a setting like Quantize, where once set, all parts will be automatically quantized on-the-fly, immediately after they've been recorded.

6.10 Advanced Quantization Options

In addition to the basic quantizing options we've covered so far, Logic also includes an expanded set of quantizing features that facilitate a more refined and delicate approach to quantizing – from the ability to "softly" quantize a

Figure 6.20
Using the MIDI Thru option as part of the Region Parameters can preset certain playback criteria – like quantizing or transposition, for example.

MIDI region, for example, through to creative possibilities like Flam quantizing. Although slightly more perplexing for novice users, Logic's Advanced Quantizing options are essential if you want to retain some of the feel of the original performance and avoid the overtly "robotic" sound of overquantized music.

You can view the Advanced Quantizing options by clicking on the small "Advanced Quantizing" arrow toward the bottom of the Region Parameters box. Once activated, you should find the Region Parameters updated to include a further set of options.

Q-Strength

Arguably, one of the most important options in the extended Region Parameters is the Q-Strength parameter. As the name suggests, this directly modifies the "gravitational" strength of quantizing so that a 100% setting, for example, pulls the notes completely to their grid position, while a setting of 50% only brings the notes halfway. This is a great way of achieving a soft quantizing – improving the overall feel of the performance, without turning it into a metronomic sequence.

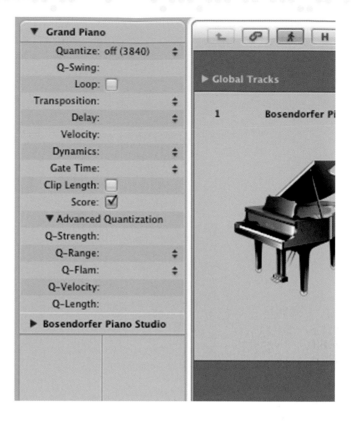

Figure 6.21 The extended Region Parameters offer more precise control over the groove and strength of quantize applied. Use these where you need to preserve important qualities of the original performance.

Q-Flam

Q-Flam avoids the sometimes-unnecessary side effect of a quantized chord having its notes aggressively bunched together. By using the Flam setting, you can "fan-out" the notes, with positive settings producing an upward movement and negative settings producing a downward movement. Although this won't be applicable to every instrument you work with, the Flam option is a feature that works really well with harpsichord or acoustic guitars samples.

Q-Range

One of the more difficult tools in the quantize tools set to use is the Q-Range control. Q-Range is another form of a soft quantizing tool, although in this case, the notion is to ignore notes closer to or on the beat, and instead direct the quantizing engine toward notes that stray beyond a given point. On a setting of zero, for example, all notes will be corrected to the current quantize settings. By decreasing the range into its negative setting, only notes that fall outside the range will be quantized, effectively creating an increasingly soft form of quantizing.

6.11 Normalizing Sequence Parameters

Should you reach a point where you're completely happy with your region's parameters, you might want to consider making the "adaptive" edits permanent – a process known as Normalizing. Although you'll lose the possibility of going back to the original performance, a normalized region might be easier to handle in the long run, especially if you manage to undo your perfected edits by mistake.

The appropriate functions can be found under the menu option MIDI > Region Parameters, with various options to Normalize Region Parameters or destructively apply any Quantize settings you've created.

6.12 MIDI Editing in Logic

Although there's a surprising amount that can be achieved using just the Region Parameters alone, eventually you hit a point where you need to explore a more detailed set of editing options.

Editors can be opened in a number of ways. For example, double-clicking on a region will open the default assigned editor (Preferences > Global, under the Editing tab), which is initially set to the Piano Roll editor. Alternatively, once one or more regions are selected in the Arrange window, you can also use the Piano Roll, Score, or Hyper Editor tabs at the bottom of the screen to open the respective editors. Finally, using the Window menu, you can also open separate editing windows away from your main arrange, which is a useful solution if you're using multiple displays.

Delving deeper still, it's also possible to view and edit your MIDI data in the form of a list of Event data – in other words, the precise note on, note off, and

Figure 6.22 Use the Normalize feature to permanently "fix" your Region Parameters like quantizing.

controller movements that your sequence is composed of. Although this might seem confusing at first, it does offer a precise and informative view of your MIDI data (much the same way Logic sees it), allowing you to position MIDI notes on a tick-by-tick basis, or set controllers to exact parameter levels.

Ultimately, good MIDI editing practice in Logic is about matching the right MIDI editor with the precise type of modification you want to achieve. For example, if your edits are generally musical, look toward the Piano Roll or Score Editors. If your edits involve a more intricate manipulation of control data, then either the Hyper Editor or List editor might be more appropriate. Not only will the right editor allow you to work quicker, but also might allow you to discover some interesting creative possibilities along the way.

6.13 The Piano Roll

The Piano Roll editor provides one of the most intuitive and informative ways of editing MIDI data in Logic and is often the first choice for the majority of MIDI editing tasks. The editor can display both MIDI note data in the top half of the interface, with the added option of viewing controller data, like pitch bend and modulation toward the bottom of the editor. Notes are displayed in much the same way as regions in the arrangement page – with the horizontal axis indicating their timing and length, while the vertical axis indicates their relative pitch. In addition to the pitch, timing, and length of the notes, you can also see their velocity, displayed as a series of colors.

Figure 6.23
Here's a typical MIDI region being edited in the Piano Roll editor, with note data toward the top of the interface and controller data toward the bottom.

Plug-In Focus 5 ▼

EXS24 Main Interface

As a completely integrated solution for using samples, Logic's EXS24 is one of its most powerful music production tools. Compared to other software samplers, like Native Instrument's Kontakt 4, the EXS24 appears a much more approachable tool, although this is largely because it makes a distinction between the front-end controls (which we're looking at here) and the behind-the-scene mapping of samples (which we'll explore in more detail in Chapter 7).

The interface is divided into a number of key sections. Most important is the Sample Instrument menu, with the current selected instrument displayed in the top part of the ESX24's interface. Clicking on this will call up the list of EXS24 Sample Instrument stored on your Mac, as well as Sample Instrument that you might have specifically created for the project in question. You can also browse through, instrument by instrument, using the small + and – buttons on either side of the Sampler Instrument menu.

Figure 6.24 Unlike other third-party samplers, EXS24 is specifically designed so that you load multiple instances – one EXS24 for each sound, or instrument, you want to appear in the mix.

(Continued)

Once a sound has been loaded in, you can then start to experiment with the various parameters settings on the interface. These include basic instrument tools – like volume, tuning, total available voices, and so on – as well as a number of synthesis-orientated features, like a multimode filter (just below the Sampler Instrument menu), three LFOs, two envelope generators, and a modulation matrix. In theory, taking the basic mapping of the samples, you can use the range of synthesis tools to transform the basic qualities of your sound, making its timbre darker, for example, or using the envelope to change the attack characteristics. We'll be taking a more detailed look at these possibilities in Chapter 7.

Unlike certain other third-party software samplers, the EXS24 isn't designed to work with multiple instruments in the same instance. Being directly part of Logic's code, the EXS24 is extremely CPU-efficient, with the intention that you run multiple instances for each sample instrument you require in the mix.

6.14 Typical Editing Scenarios in the Piano Roll

If you're used to editing regions in the Arrange window, you should find that many of the techniques carry directly over to the editing in the Piano Roll editor. As with the Arrangement window, this editing process is somewhat tied up with a series of tools, including familiar objects (like the pointer tool, pencil tool, and so on) alongside two tools that are unique to the Piano Roll – the quantize tool and the velocity tool.

To understand how the various tools work and relate to each other, let's look at a variety of editing scenarios and how you might combine the various tools and techniques in the Piano Roll editor to rectify this situation.

Correcting Pitch

Using the pointer tool, you can move a note vertically up or down to change its relative pitch. By adding in a few keyboard modifiers, you can also perform some quick modifications without needing to drag objects about. For example, press Alt + either the up or down arrows and you can move a note or selection of notes up or down in semitone steps. Add in shift (Alt + Shift + Up/Down) and you can move the same notes in octaves.

Correcting Timing

Although quantizing can offer a quick-and-easy route to correcting the timing problems of a particular MIDI sequence, sometimes it can be important to have a more precise control over the placement of notes in your sequence. By

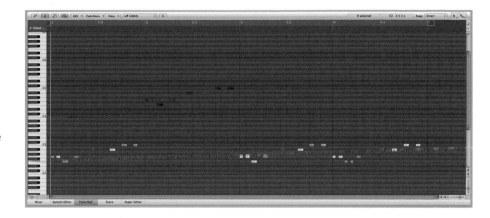

Figure 6.25 Use Alt + Shift and either of the up or down keys to move selected notes in octaves.

leaving the region unquantized, you can use the pointer tool to manually position single notes, or groups of notes, by hand.

As with audio editing, it's important to note the use and operation of the Snap feature (found in the top right-hand corner of the Piano Roll) and how this interacts with the movement of the notes on the Piano Roll's grid. In most cases, this is set to its Smart setting, allowing you to place the notes anywhere along the timeline, although with a definable "pull" to the grid points. If you specifically want the notes to always snap to the grid, you can change this to beat or division, for example, depending on how fine you want the resultant grid. For truly precise movements, though, it's worth adding in a few keyboard modifiers. Shift + Ctrl will switch off any grid setting – irrespective of the current Snap setting – allowing you to precisely move the note in question. Alternatively, use the Nudge feature (Alt + the left or right arrow) to move the note tick-by-tick using the keyboard.

If you want to get creative, it's also worth experimenting with the timing of phrases by shifting them about in various divisions. For example, try taking an existing phrase and moving this by 1/16th, 2/16ths, or 3/16ths. Notice how the feel of the phrase has changed with the notes falling in a different placement. You can also use this to good effect by starting a track with a phrase that doesn't fall on the first downbeat of the bar, and then, as the drums come in, the listener experiences mental "shift" as they're forced to relocate the bar.

As you'd expect, there's also a sophisticated range of quantizing tools available from within the Piano Roll editor. This will be discussed later in Section 6.15.

Duplicating a Phrase

By adding in the Alt modifier as you drag your selection, you can create a copy of the note(s) in question. This is an effective way of duplicating musical material within the region.

Correcting Length Issues

The exact length of a note can be as important to the groove of a part as the note's start point. The specific tool assigned to this task is the finger tool, although you can achieve a similar result using the pencil tool and pointer tools by clicking toward the end of a note's duration. Try selecting one or more notes and reducing their lengths to create a sharp, staccato-like playing style. Alternatively, try lengthening the notes to produce a smoother, more legato-like effect. Mixing and matching these different length settings can be surprisingly effective on basslines.

Changing Velocity

Velocity – or in other words, how hard you play a note – has a big part to play in the expression in your music. Ideally, the best types of music incorporate a range of different velocities – both from note to note and over the whole "emotional arc" of the piece. The velocity tool in the Piano Roll editor allows you to change the velocity of any note or group of notes you click on. Simply click on the note and move the mouse up or down accordingly to increase or decrease the current velocity accordingly. Where a group of notes is selected, the velocity will keep their respective balance (in other words, some notes louder than other).

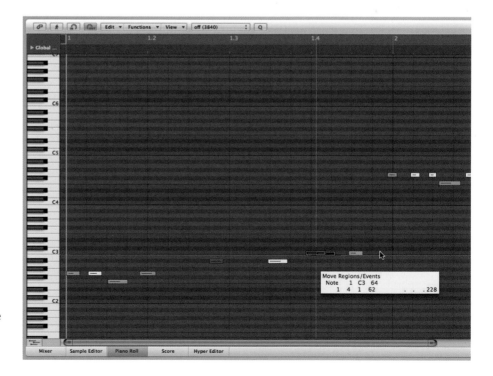

Figure 6.26
Holding Shift + Ctrl as you move a note allows you to negate any preexisting Snap settings and place each note to a precise tick-based position.

One aspect of editing where velocity really comes in important is drum programming. As you're putting together a pattern, be sure to experiment with different accents on the various main beats, as well as adding discrete ghost notes using lower velocity settings.

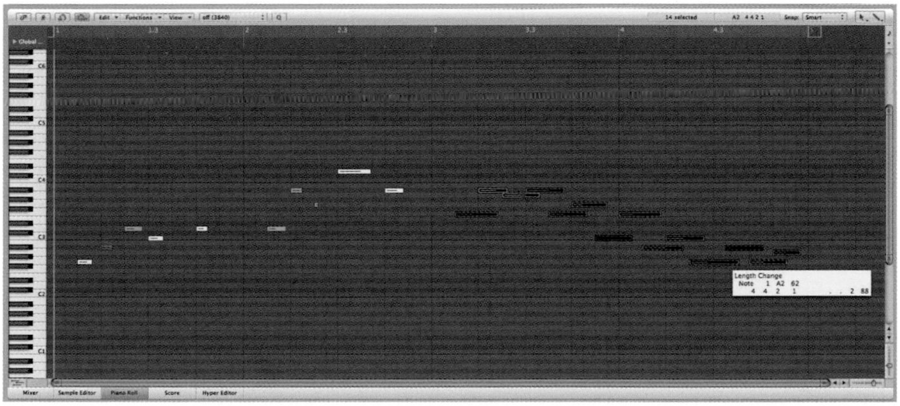

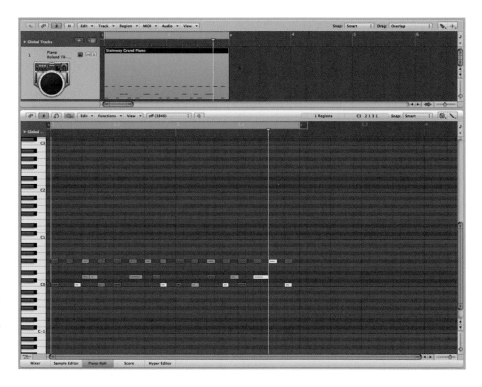

Figure 6.27 Use the finder tool to adjust the length of one or more selected notes – moving between a staccato and legato playing style.

Figure 6.28 Here's an example of a programmed drum loop making extensive use of the velocity tool to define the accent and grove of the loop.

Although the velocity tool provides one solution to changing a note's loudness, it's also possible to use the Hyper Draw controllers display to also achieve the same task. Have a look at Section 6.16 for more information.

Creating New Notes

Use the pencil tool as a means of creating new notes in your region. As we saw in the section on changing length issues, the pencil tool can also be used to define the note length by clicking toward the end of the block. One other trick to look out for is how the pencil defaults to creating a note based on the last edited note event. For example, if you create a note and modify to be 1/64th long with a low velocity, all subsequent notes created with the pencil tool will conform to that standard. This can be a useful way of setting a default note type for all subsequent newly created notes to follow.

Knowledgebase 1 ▼

EXS24 Virtual Memory and Disk Streaming

In an attempt to make a realistic recreation of acoustic instruments, many third-party sample libraries can make quite exhaustive demands on your RAM. For example, some Piano instruments can comprise multiple velocity layers, with a separate set of samples for each key on an 88-note piano keyboard. It's not uncommon, therefore, for these instruments to have RAM demands between 1 and 2 GB – something that many users' systems will fail to cope with.

If you do run out of available RAM (the sampler will tell you that it has been exceeded and will fail to achieve a complete load), you might need to consider using the EXS24's Virtual Memory feature. Virtual Memory works by only loading a small amount of the start of each sample, with the remainder streamed from your hard drive as and when required. By using Virtual Memory, you'll be able to load a far greater number of these large instruments into a project and have them played back successfully.

To access the Virtual Memory feature, open up the EXS24 Options menu and select Virtual Memory from the list. A dialog box will allow you to set and optimize settings, balancing off the hard drive's speed with the amount of disk activity (namely, audio track streaming and recording) also happening at the same time. Obviously, if you're already running a lot of audio tracks, this might begin to put your hard drive under an increasing amount of strain, whereas a low amount of disk activity will ensure optimal performance with a large amount of samples voices being playable

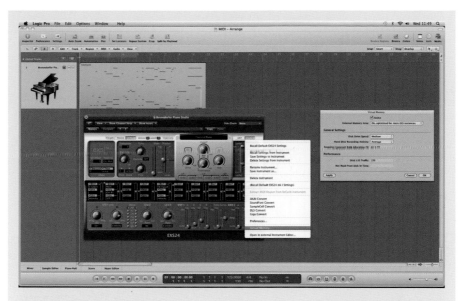

Figure 6.29 With Virtual Memory engaged, you'll be able to load a greater number of RAM-hungry software instruments. Be aware, though, of potential knock-on effects on your system's performance.

at the same time. You might also want to consider using an external Fire 800 drive, or an additional Serial Advanced Technology Attachment (SATA) drive (internal or external), to keep certain types of data accessible quickly and easily.

In most cases, the usual "Core Audio Overload" will indicate the system being pushed too hard, although you might also want to listen out for longer notes tails, which can get snatched off by excessive disk activity. Also check out the Disk I/O Traffic and Not Reading from Disk in Time performance indicators in the Virtual Memory Dialog box. These will also indicate the effective, or problematic, use of your system resources.

Deleting/Muting Notes

To remove a note from a region, use the delete tool or simply select the required notes and hit delete on your keyboard. However, a more strategic solution might be to use the mute tool. By muting the note event, it will cease to be played back from the region, but at any point, you can return to the region and unmute the event to hear the original note as it was once played.

One other useful application of the muted notes feature is as an editing guide. For example, say you're trying to harmonize or augment an existing sequence

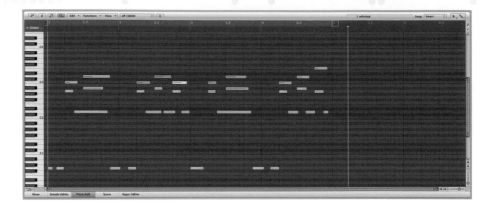

Figure 6.30
Muting notes is a useful alternative to delete them. In this example, a muted line has been used as a guide to create a harmonized part.

using a different virtual instrument. By copying the region and muting the original notes, you can visualize what the other part is playing, while at the same time constructing or editing a new sequence to play over the top.

6.15 Quantizing Inside the Piano Roll Editor

Despite the Inspector being the first port of call for quantizing tasks in Logic, it's also interesting to note that another layer of quantizing exists within the Piano Roll editor itself. As flexible as this is, this two-tier system can appear somewhat confusing at first, so it's well worth spending a few minutes familiarizing yourself with the interaction between these two layers of quantizing.

Start by recording a simple sequence into Logic and open this up in the Piano Roll editor. At this point, you should see the various imprecise placements of the notes to match what you originally played. Now move up to the Region Parameters, as part of the Inspector, and set in Quantize setting to an appropriate value, like 1/16th. Glancing back at the edited part, you should now see the notes snapped exactly to the grid.

Now let's try quantizing from directly inside the editor itself. One technique for quantizing is to select one or more notes in the Piano Roll editor and then pick a desirable quantize setting (like 1/16 note) from the drop-down Quantize menu in the editor window itself. Another technique is to select the desired quantize setting first, then use the quantize tool to nudge the appropriate notes into line (this is a good way of being slightly more selective with your quantizing). Note, however, that with both techniques the settings with the Piano Roll editor ignores any settings made in the Region Parameters. However, should any settings be reapplied in as part of the Region Parameters and then the current quantize settings (in the Piano Roll) are lost.

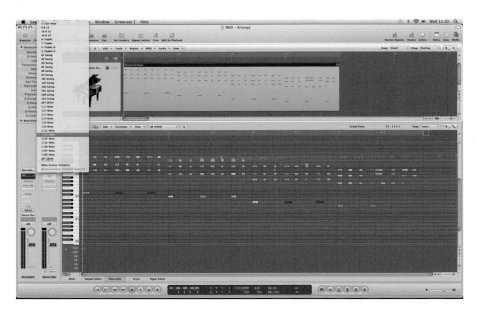

Figure 6.31 Quantize settings made in the Region Parameters box will automatically be visualized in the Piano Roll editor.

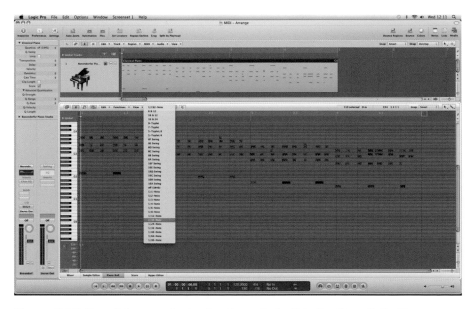

Figure 6.32 Quantizing can also be directly applied within the Piano Roll editor, negating any exiting quantize settings that have already been set in the Region Parameters box. This allows you to quantize on a note-by-note basis.

6.16 Working with Controller Data Using Hyper Draw

Although the Piano Roll editor does an excellent job with note data, it doesn't quite provide the full picture with respect to MIDI controller data – including dynamic movements of modulation, pitch bend, volume, and pan data. However, by opening up the Hyper Draw view at the bottom of the Piano Roll window, you'll be able to see controller data as a series of lines with nodes along their length to edit the respective parameter levels. To open the Hyper Draw, view select View > Hyper Draw (followed by the controller of your choice from the list) or open the window using the tab in the bottom left-hand corner of the Piano Roll window.

To better understand what can be achieved in the Piano Roll editor, let's take a look at some typical editing tasks.

Editing Velocity in Hyper Draw

Although we can edit velocity data in the conventional note display of the Piano Roll editor, there are several operations – like a gradual rise or decay of velocity, for example – that can be better achieved using Hyper Draw. To edit velocity using Hyper Draw, you'll need to select the Velocity option from the drop-down list accessible from the small arrow icon to the left of the Hyper Draw area. Once the Velocity view is active, you should be able to adjust the individual velocities, although its most powerful feature is creating a gradual fall or rise over time. To do this, click and hold with the mouse, only releasing when you find a suitable "start point" for the line. Now, move the pointer over to end point, noticing the trail that is left behind it. When you click again, the virtual line will be drawn, effectively raising or lowering the velocities accordingly.

Figure 6.33
Editing velocity using the hyper view is a much better way of drawing in a gradual rise or fall of velocity.

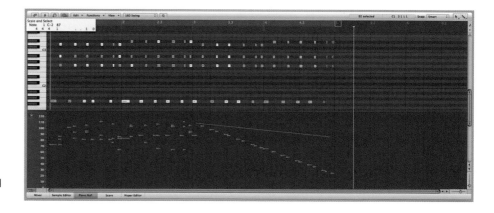

Editing Volume

Volume moves – like a number of other generic controllers including pan, modulation, and pitch bend – can also be drawn directly into the Hyper Draw area, allowing you to creatively shape an instrument's sound proprieties over time. In the specific case of volume and pan (assuming you're using it to control a software instrument), you'll actually see the volume and pan controls on the mixer move to reflect the curves you've drawn in the hyper view.

Drawing curves for these controllers can be done in two ways. The first technique is to use the pencil and effectively draw in the moves "freehand," although, in fact, Logic will actually place a series of node points along the duration of the part. The second technique is the pointer tool that allows you to place, edit, or remove the nodes one by one, which might be a better solution for more controlled movements.

Editing Other Controllers, Including Filter Cutoff

If you can't see the controller that you need in the pop-up list, you can always use the Other option to access any 1 of the 127 different MIDI controller message types. For example, it might be that your software instrument or hardware synthesizer responds to MIDI controller message as a means of controlling filter cutoff, for example, or that it can be configured to work with certain controller message.

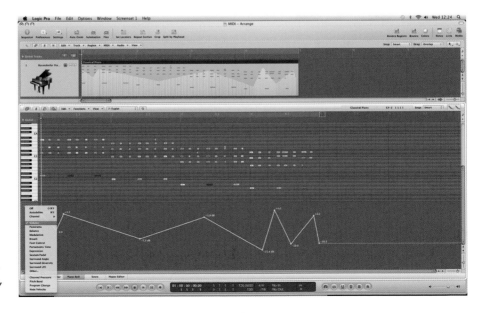

Figure 6.34
Hyper Draw can provide an effective solution for drawing in more generic controller moves – like volume or pan, for example.

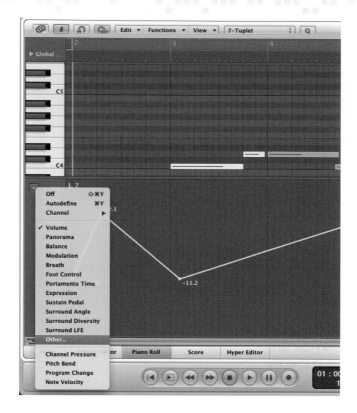

Figure 6.35 Use the Other option to display a MIDI controller type outside the main functions like volume and pan.

6.17 Hyper Draw in the Arrange Area

Although we're going to hop out of the Piano Roll editor for now, it's important to note the parity between Hyper Draw editing in the Piano Roll and the Hyper Draw facilities in the Arrange area. In effect, the Arrange area provides the full range of Hyper Draw features that we find in the Piano Roll editor all conveniently accessible via the View menu. Immediately, you should notice small previews of the moves we've made, but by selecting a region, choosing View > Hyper Draw, and selecting the controller type from the drop-down menu, we can see and edit the controller moves as if we were in the Piano Roll editor itself.

As you'll see, the Hyper Draw feature is very close to the functionality of Logic's Automation system – something that we'll take a closer look at in Chapter 8. In particular, what we're playing here is a form of region-based automation – in other words, data and automation moves that are directly stored within the MIDI data of a region.

6.18 Going Further: The Piano Roll's Edit and Functions Menus

One of the quickest ways of speeding up the editing process is to explore alternative methods of selecting and processing note events. For example, you might only be interested in notes after a given position in the region, notes that have been previously muted, or notes at the top or bottom of a group of chords. By applying an automated selection process, you can achieve complex editing operations with relative speed and efficiency. Equally, various features in the Functions menu allow you to select and manipulate MIDI information in ways you wouldn't immediately think of, but is, nevertheless, a gold mine of useful creative options and technical solutions to MIDI editing.

6.19 Intelligent Selection: The Edit Menu

Besides the usual cut/copy and paste options, the Edit menu contains a number of different selection criteria. To better understand the Edit menu functions in the production process, though, let's look at some key musical applications of these features.

Select Equal Subpositions is a great tool for working with any rhythmic regions where the material spans several bars. On a hi-hat pattern, for example, we might want to accent particular beats across a number of consecutive bars. Start by selecting the appropriate notes in the first bar of the sequence and then go to the Edit > Select Equal Subpositions. Having activated the function, you should now see the appropriate notes selected, which you can then increase the velocity of using the velocity tool.

Invert Selection is a useful way of "toggling" selected notes in the Piano Roll editor. Say, for example, you're interested in modifying a large number of notes (maybe the unaccented notes in the example of the hi-hat), but at the same

Figure 6.37 The Edit and Functions menu facilitates powerful tools both for the selection and processing of MIDI data, making MIDI editing both more efficient and more creative.

time, leaving a few notes unmodified. In this example, you'll achieve the results much quicker by selecting the notes that you don't want to modify, then inverting the selection (Shift + I) to toggle the selection, leaving the actual notes you want to modify selected.

Invert Selection is a great editing function to use in conjunction with the note-muting feature. For example, say you want to distribute note data between two virtual instruments from the same "parent" region. Starting with the first region, use the mute tool to remove the notes that you aren't interested in hearing with this particular instrument. Now copy the region over to another instrument track, open it up, and select the muted notes using the Edit > Select Muted Notes. Unmute the selected notes (M), invert the selection (Shift + I), and then mute the newly selected notes. From one simple selection, you should have now made two regions with an alternate set of active notes.

Figure 6.38 Here's our simple hi-hat pattern; in the first bar, we've selected the accent points that we'd like to pick up across all the bars in the region.

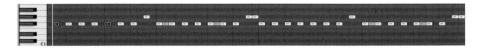

Figure 6.39 Using the Select Equal Subposition, we can add the corresponding notes in the subsequent bars quickly and easily. Now, all we need to do is raise the velocity to complete the effect.

Figure 6.40
Inverting and muting selections between different instruments can be a useful way of distributing MIDI information across your arrangement.

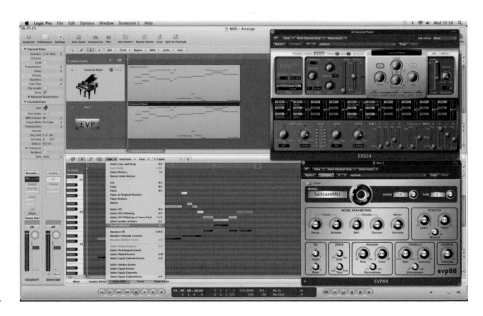

Select Equal Events

Now and again it's possible for two MIDI events, of the same pitch, to end up on top of each other. This usually happens as a by-product of quantization, and, if you listen carefully, can be heard as a slightly "airiness" or "phasing" on the note in question. If you suspect this is happening, one quick way of rectifying this is to use the Select Equal Events option from the Edit menu and then hit delete on your keyboard. Although you might not see any immediate difference, Logic will have deleted these "double" events, with the note sounding a more natural way.

6.20 Functions: Quick-and-Easy Note Modifications

The Functions menu includes a number of different tools for both selecting notes and applying various transformations to MIDI data. The tools include some marvelous tools for dealing with music arrangement, especially with orchestral samples – either to create fluid legato passages, for example, or split played chords out into a series of parts.

Note Overlap Correction/Note Force Legato (Functions > Note Events)

Both these functions are interesting tools to work with when you're working with instrument samples – like violin or clarinet, for example – although they can have a number of musical applications. Use Note Overlap Correction with a MIDI region being directed to a "monophonic" acoustic instrument, like an oboe or clarinet. In the real world, it is impossible for these instruments to play two notes at the same time, so any accidental note overruns (where one note has been held for too long) can affect the realism of the output. Note Overlap Correction, therefore, will remove any accidental overlapping of notes.

Alternatively, Note Force Legato is a great solution where you want the part to take on a fluid, expressive quality – like violins playing a lead line. With Note Force Legato, all the notes run into each other, irrespective of their original played length, producing a smoother overall line.

Select Highest Notes, Select Lowest Notes (Functions > Note Events)

By using the keyboard shortcut Shift + Up arrow or Shift + Down arrow, you can select the top and bottom line of a polyphonic region (in other words, a region with multiple notes at any given point in time). One immediate application of this is dividing a played chord of the keyboard between three or more monophonic string instruments (violin, viola, and cello, for example). To do

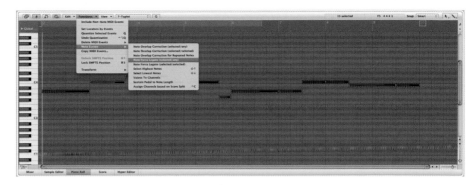

Figure 6.41 Using the Note Force Legato feature (Functions > Note Events > Note Force Legato), this MIDI region has had all its notes lengthened to meet the subsequent note.

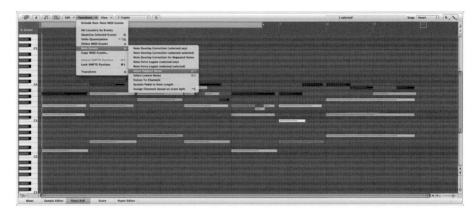

Figure 6.42 Use the Select Highest Notes or Select Lowest Notes to strip out the top or bottom line of a polyphonic MIDI performance. This could then be assigned to different instruments, for example.

this, simply copy the three-part harmony over to the various instruments and use the Select Highest Notes, Select Lowest Notes accordingly, along with the mute function, to remove the required notes in each part.

Voices to Channels (Functions > Note Events)

Use the Voice to Channels feature as an intelligent, automatic means of splitting a polyphonic chord part into a series of unique voices. This end region will look the same, although on closer inspection you'll find that each voice has been assigned a different MIDI channel. To strip the data out, select the region and choose MIDI > Separate MIDI Events > By Event Channel. The polyphonic region should now be exploded into a number of different regions, one for each "voice" of the original part.

Knowledgebase 2 ▼

Using Multioutput Instruments

Both the EXS24 and Ultrabeat instruments are available in multioutput versions, in addition to their standard mono and stereo configurations. It's also possible that a number of other third-party software instruments in your collection – like FXpansion's BFD – might also come in multioutput versions. The notion of a multioutput is simply to provide additional outputs from the plug-in. Each output has its own fader, and, therefore, the possibility of adding additional effect plug-ins specific to the output in question, rather than the instrument as a whole. This is most often the case with drums, where you might wish to place different reverb, EQ, and compression settings on each different part of the kit.

To access the additional output, you'll need to ensure that you instantiate the specific multioutput version of the plug-in. To add in additional aux channels, simply click on the 1 sign at the bottom of the instrument channel strip, with each subsequent press adding another output into the equation. To access the outputs, you'll need to route the appropriate signals using the plug-in's controls. On the EXS24, for example, the routing can be achieved via the EXS24 editor, while Ultrabeat's output options can be found next to each drum voice.

Figure 6.43 Use the multioutput versions of instruments like the EXS24 and Ultrabeat so as to be able to process and mix individual elements from their sound output.

6.21 Step-Time Sequencing

Step-time sequencing dates back to the earliest analog CV sequencers that wouldn't technically allow you to record real-time musical information (as Logic does), but instead store music as a series of 16–32 control voltages. Nowadays, however, step-time sequencing is still a valuable tool, mainly as a means of "stepping through" and creating a MIDI sequence piece-by-piece, rather than having to play a part by hand. If you're creating sequencer-driven music – with flowing arpeggiators and fast-moving basslines – then step-time sequencing arguably offers a more effective means of recording a part, without the need to quantize and meticulously edit your original sloppy performance.

To engage step-time recording in the Piano Roll editor, you'll need to activate the small MIDI In icon in the top left-hand corner of the window. Whenever you press a MIDI note in step-time mode, Logic will place one or more (if you play a chord) notes at the current song position and then advance one step forward at an amount defined by the current Snap setting. Working in this way, you can quickly create a fast-moving sequencer line playing each note in sequence, but without having to play in time with the metronome.

As an alternative to using a MIDI keyboard, you can also make use of Logic's Caps Lock Keyboard (a feature that allows to play an instrument using your Qwerty keyboard, which can be called up by pressing Caps Lock down) as well as the dedicated Step Input Keyboard found under Options > Step Input Keyboard. The Step Input Keyboard is a good solution for drawing note events

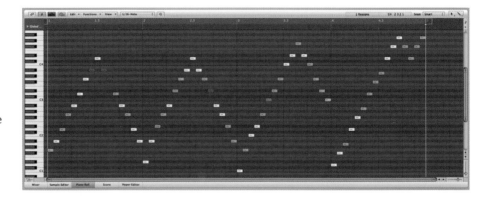

Figure 6.44 Here's an example of sequence created in step time, with a fast-moving sixteenth arpeggio effect.

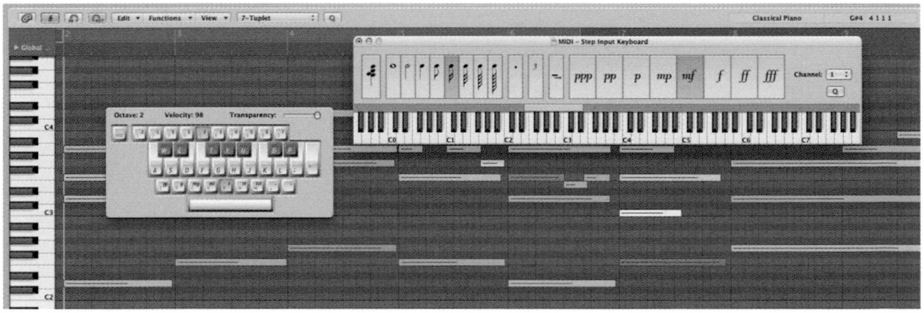

Figure 6.45 Use the Step Input Keyboard and Caps Lock Keyboard as alternative methods of note entry for step-time recording.

using the mouse, with a clear horizontal representation of the musical keyboard, selectable note lengths, velocity settings, and so on.

6.22 The Hyper Editor

Although we've spent some detailed time exploring Logic's main MIDI editor – the Piano Roll – it's worth familiarizing yourself with some of the other MIDI editors and what they can individually offer to the MIDI editing process. Having seen Hyper Draw, therefore, the specific Hyper Editor shouldn't come as a great surprise. As with Hyper Draw, the main focus of the Hyper Editor is in the editing of MIDI controller data – like volume, pan, or modulation – although the specific benefit here is that we get to see a number of track lanes at the same time.

Editing and Creating MIDI Controller Data in the Hyper Editor

One quality that should become apparent about the Hyper Editor is that it prefers to display its controller data as a series of bar graphs (usually on a six-teenth grid), rather than the line + node approach of Hyper Draw. Technically, this creates a slightly different approach to editing, although the end effect is much the same. It does, however, also make for an excellent source of pseu-dorandom sixteenth Sample-and-Hold movement, where a pattern appears to

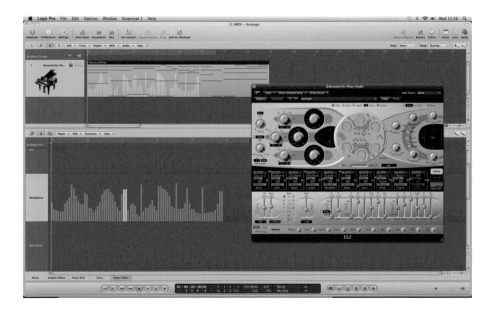

Figure 6.46 The Hyper Editor is particularly good at stepped sixteenth controller movements.

pulse up and down in time with the music. Try doing this with the modulation controller, for example, mapped to filter cutoff on the ES2 synthesizer.

For more generalized movements, you can use the pencil tool to draw in a series of stepped MIDI events – maybe creating a quick ramping up and down of volume, for example. Using the line tool, you can also draw more gradual movements up and down, in the same way as the line feature in Hyper Draw.

Knowledgebase 3 ▽

Separating MIDI Events

The Separate MIDI Events feature is a simple way of exploding a MIDI region into its constituent parts – this could be based on note pitch, for example, or by the Event Channel. To Demix a region, first select it, then choose MIDI > Separate MIDI Events > By Event Channel or MIDI > Separate MIDI Events > By Note Pitch. The note pitch option, for example, is a good solution to use in combination with a MIDI drum loop, where a number of different lines (hi-hat, Toms, and so on) have been sequenced into the same region. By separating the MIDI region, you'll create individual parts for each part of the kit, which you can then label Kick Drum, or Snare, for example, accordingly.

The Event Channel option works well combined with the Functions > Note Events > Voices to Channels feature in the Piano Roll, where a polyphonic part has been split into a number of different MIDI voices.

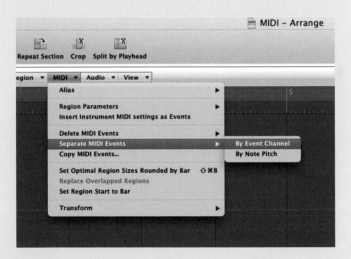

Figure 6.47 Demixing allows you to explode a single region into a number of separate parts based on note pitch or MIDI channels.

Adding New Controller Types and Hyper Sets

To add a new controller type other than the ones in the preexisting set, simply double-click in the blank space beneath the last controller assignment, effectively creating a new controller track lane. Opening up the Inspector, you can then define parameters in relation to the controller lane, including the pen width, grid, and most importantly, the MIDI status type (program, controller, and so on), controller number (if applicable), and MIDI channel.

The particular arrangement and order of controller track lanes can be saved of as part of your own, custom-designed Hyper set. You could, for example, create a Hyper set that specifically relates to the MIDI control of a particular instrument plug-in or a simplified set based on mixing. To create or delete a Hyper

Figure 6.48
Use the GM Drum Kit Hyper Set as a useful way of creating and editing drum sequences. Note that you can also create your own Hyper sets to match the mapping of virtual instruments like BFD.

set, go to the Hyper menu and select the corresponding option or alternatively use the dedicated Hyper set selector (where you can also switch sets) embedded into the Inspector.

Editing Drums Using the Hyper Editor

In addition to editing controller data, the Hyper Editor can also be used to edit and create note data, using the same series of sixteenth "blocks." As such, the Hyper Editor is a good solution for drum programming, with the height of each block providing quick and easy velocity adjustment, while the pencil tool allowing to draw in straight lines of sixteenth events. Usefully, Logic provides a precreated GM Drum Kit Hyper Set, specifically designed for drum programming. As most drum kits are mapped to the GM standard, this should be good for the majority of applications, although you can always program your own drum-based Hyper sets for plug-ins like BFD or Battery.

6.23 Score Editor

The Score Editor is the final MIDI editor selectable from the bottom row of tabs. Obviously, this editor is an appropriate choice if you have an ability to read printed music, although it has several key limitations for serious and detailed MIDI editing tasks. As a score is effectively "visually" quantized, the Score Editor provides us with a relative coarse view of a note's placement and, while we're at it, its length. This makes any change beyond pitch manipulation somewhat flawed.

Of course, the real benefit in including the Score Editor is in the preparation of music for musicians to play. Music preparation, as it's called, is a process we'll be taking a closer look at in Chapter 11.

6.24 Event List

The last MIDI editor that we're going to take a look at is the Event List editor, which displays the MIDI information in pure "data" form. The Event List can, at first, feel a little daunting and perplexing, and certainly not comparable to the visually driven approach to MIDI editing defined in the Arrange area, Piano Roll, Score, and Hyper Editor. However, there are times when it's important to get "down-and-dirty" with the data – understanding precisely where an event occurs, for example (down to tick-level accuracy), or the exact parameter level with a controller.

The Event List editor can be opened as part of Logic general "list" displays – including markers, tempo, and time signature lists – that can be opened up on the right-hand side of the Arrange window. This is an effective solution, as

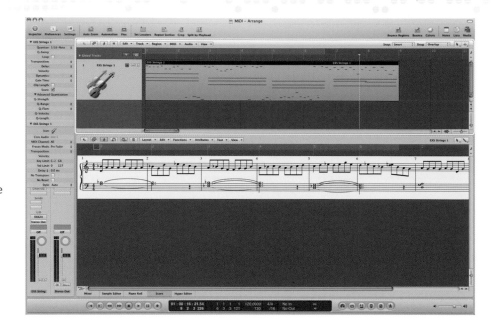

Figure 6.49 The Score Editor is a quick-and-easy way of editing pitch-based information, but provides us with little control or information on positional data.

Figure 6.50 The Event List provides an alternative text-based view of our MIDI data. Combined with the other editors, the Event List can be an indispensable tool to accurately understand the content of your MIDI region.

often, you'll want to combine the Event List with other MIDI editors like the Piano Roll or Hyper Editor. To open the Lists display, press E on your keyboard, or View > Lists, with the Event List accessible from the Event tab at the top of the display.

Filtering Events

Given the range and quantity of MIDI data that can be included in a region, it is often necessary to filter different types of MIDI data out of the Event List. By clicking on the Filter tab, you can then switch in and out various MIDI events, including notes, program change, controller, and so on. Using the filter can also be an interesting way of selectively deleting certain types of MIDI data. For example, you could filter note data so as to select and then delete extraneous MIDI data (controllers and so on) that you might not be interested in.

Modifying and Adding Events

Editing data in the Event List is as simple as clicking on the appropriate number and either dragging up and down or double-clicking and entering a new value

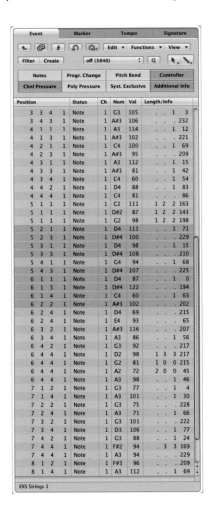

Figure 6.51 Use the Filter tab as means of filtering out certain types of MIDI data.

numerically. Numerical entry can be a surprisingly effective way of positioning notes and other MIDI events both quickly and accurately. You can also add new events, based on the current song position, by clicking on the Create tab and then selecting an accompanying message type. For program changes, or static controller levels, the Event List shines as a quick effective solution with the minimum of fuss.

Knowledgebase 4 ▼

EXS24 Data Management

Although it's quite possible to operate and work with Logic without an understanding of the EXS24's occasionally complex means of data management, it can become increasingly important as your own user-installed collection of sampler instruments begins to swell. The key to understanding where your data is, and how this is managed, is in differentiating the different sources of sample instruments – in other words, original library data installed with Logic, user-installed and created content, and data that specifically relates to the current project. Although this can seem complicated at first, a little time investment will allow you to better access, organize, and archive this important data.

Factory content that was installed with Logic can be found under the path Library/Application Support/Logic/Sampler Instruments, with the associated sample files (remember an instrument file is separate to the sample data) under Library/Application Support/Logic/EXS Factory Samples. Your data will be saved in your user file under User Name/Library/Application Support/Logic/Sampler Instruments. Whenever Logic is loaded, it will scan these two folders and build up an Instrument list based on the contents. This also means that you can actively use the Finder to organize the instrument files as you see fit, creating custom list of folders, for example, or even moving or duplicating favorite instruments into a single folder.

In addition to these two main folders, Logic will also take a look inside your Song folder to see if any EXS instruments are stored there. This may well be the case if you've inherited a project from another user who's backed-up a complete set of assets with their project archive. It should be pointed out, though, that these project-specific instruments can't be accessed in other projects. To do this, you'll need to copy the instrument files into the appropriate Library folder.

Samples for any of these instruments can reside on any drive, in any folder, with Logic scanning for the appropriate samples as it loads the instruments. The two immediate implications here, though, are both the scattershot sample organization that can occur and the speed implication of a fragmented collection of samples. If possible, therefore, try to organize a large part of your sample data (you can leave the EXS Factory data) into a neatly separated folder, partition, or even better, a separate drive.

Figure 6.52 You can use the Finder to actively manage the organization and arrangement of your sampler instruments and how these appear in the EXS24 instrument menu.

In This Chapter

7.1 Introduction 199

7.2 Logic's Synthesizers 199

7.3 Understanding the ES2 202

7.4 Working with Oscillators 205

7.5 Filters, Amplifiers, and
Modulation 210

7.6 Global Parameters and Output
Effects 216

7.7 EFM1 and Frequency Modulation
Synthesis 216

7.8 Component Modeling:
Sculpture 222

7.9 Objects 223

7.10 The String 226

7.11 Waveshapers and Beyond 229

7.12 Modulation and Morphing 232

7.13 Creative Sampling 237

7.14 The EXS24 Instrument
Editor 239

7.15 Creating a New Instrument and
Importing Samples 242

7.16 Changing Zone Properties 244

7.17 Working with the EXS24's
Groups 248

7.18 Editing EXS24 Instruments Using
the Front Panel 250

7.19 Ultrabeat 254

Knowledgebases

Waveshapes 203

Oscillator Sync 208

Filter Types 210

Vector Synthesis 214

ES2 Filter Routing 217

Wavetable Synthesis 221

Bounce in Place 229

Creating EXS Instruments from Audio
Regions 238

ReCycle Files and the EXS24 240

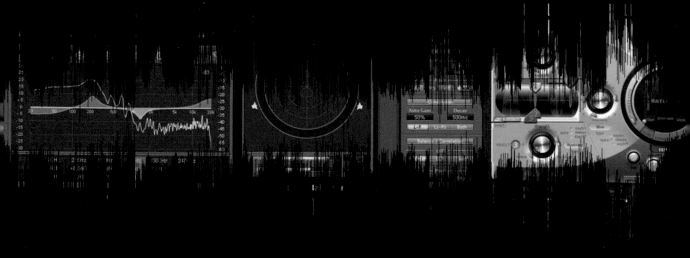

Creative Sound Design

7.1 Introduction

For anybody actively engaged in the more creative aspects of designing and modifying sound – say, a film sound designer, for example, or a musician working in the realms of electronica – there can be little doubt that Logic represents one of the most exciting applications to work in. Right out of the box, the modern-day sound designer has access to a huge range or synthesizers, samplers, and signal processors without having to go anywhere near any third-party products. Indeed, it's highly likely that your whole sound design project (whatever form it takes) could be realistically achieved entirely within the realms of Logic.

Having worked on the basics of using virtual instruments and MIDI in the last chapter, this chapter goes on to look at the creative applications of Logic's virtual technology to shape and craft any number of interesting sounds – whether you're using its versatile set of synthesizers (ES2, Sculpture, and so on), the pure sonic muscle of the EXS24, or the range of esoteric signal processors. Most important, however, making your own sounds (rather than just simply relying on presets) can be a real opportunity to stamp your own identity onto a particular audio product, and ultimately, make your work stand out from the crowd!

7.2 Logic's Synthesizers

Beyond some of the basic synthesizers we looked at in Chapter 6 – the ES E, ES P, and the ES M synthesizers – Logic also includes a number of other synthesizer plug-ins, each dedicated to a different branch of synthesis and capable of an enormously wide palate of potential sounds. So, let's take a look at the three remaining synthesizer plug-ins, and what particular strengths they offer to the sound designer.

DOI: 10.1016/B978-0-240-52193-0.00007-1

ES2

The ES2 is Logic's definitive answer to the good-old polyphonic analog subtractive synth – like Roland's Jupiter 8, for example, or Sequential Circuits' Prophet 5. However, this only forms part of the picture, as the ES2 has plenty of further tricks up its sleeve, namely, vector synthesis, wavetable synthesis, and a host of other sound design tricks borrowed from many of the classic digital synths of our time: the Waldorf's MicroWave, the Korg Wavestation, and the Prophet VS. Make no mistake, this a tremendously versatile synth – not the easiest of plug-ins to get your head around, but one that will consistently surprise you with what it has to offer.

Figure 7.1
The ES2 – arguably Logic's most powerful and sonic dexterous synthesizer.

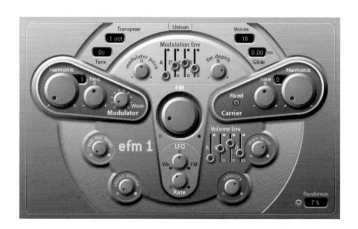

Figure 7.2
The EFM1 is a simple synthesizer designed to exploit the digital wonders of FM synthesis.

EFM1

The EFM1, although technically part of Logic's more budget-orientated synths, is actually an interesting and valid example of FM synthesis in action. For those of you who remember, FM synthesis was the primary system for synthesis used by Yamaha in the 1980s, and was at the heart of their infamous DX7. Although technology has moved on significantly since the golden days of FM, it can still be a valid source of sounds impossible to achieve elsewhere.

Sculpture

Sculpture is easily the most cutting-edge synthesizer in Logic's collection, and represents one of the most exciting developments in synthesis in the recent years – that of Component Modeling. Unlike synths like the ES2 and EFM1, Sculpture works on the basis of a series of complex mathematical models describing the precise acoustic behavior of a particular instrument. For example, rather than synthesizing the sound of a guitar by selecting a "similar" waveshape and approximating its behavior with filters and envelopes, Sculpture precisely understands what happens as a string is plucked (using a variety of different plucking mechanisms) and what happens as the sound dies away. Not surprisingly, therefore, Sculpture produces sounds impossible to achieve on other synthesizers.

Figure 7.3
Sculpture is a unique software synthesizer dedicated to Component Modelling synthesis.

7.3 Understanding the ES2

For most conventional music and sound design activities, the ES2 will probably be your main synthesizer of choice – both for the simplicity of working its interface, alongside the range of subtractive synthesizer sounds so eminently useable in contemporary electronic productions. The full interface itself is divided into the five key areas, illustrating the principal audio signal flow through the synthesizer (Oscillator > Filter > Amplifier, from left to right) alongside the interaction of modulation sources like low-frequency oscillators (LFOs) and ADSR (Attack, Decay, Sustain, and Release) envelope generators. An additional "Macro only" mode – selectable from the bottom of the ES2 – presents a minimized control panel, which might be suitable if you're only interested in tweaking presets.

The Oscillators (A)

Toward the top left-hand corner to the interface, you'll find the controls for the three oscillators included within the ES2. The oscillators form the starting point of sounds created on the ES2 – the raw, harmonically rich waveshapes from which the filter gauges out harmonics, and a big part of the identity of the

Figure 7.4

sound. Thankfully, the ES2 includes a number of different principal waveshapes, and even a wavetable (more on this later) as well as a unique triangular vector mixer for the purpose of blending the oscillators together.

The Filters (B)

Once the waveshapes have been selected and mixed, they are passed on to the circular filter section. As previously mentioned, the filters subtract harmonics from the original waveshapes – making the sound darker or thinner depending on the exact filter settings. On a musical level, the filters are important for creating the "juicy" sound of analog synthesis, as best exemplified in sounds like the classic TB-303 bassline used in dance music. However, what's unique about the ES2's filter section is the fact that it contains two completely independent filters, opening up any number of different effects only possible with fully modular synthesizers.

Knowledgebase 1 ▼

Waveshapes

As you've seen with the oscillator section of the ES2, waveshapes can have a big effect on the type of sound you can produce. However, what defines the sound of each waveshape, and how best can they be used?

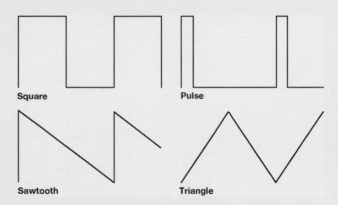

Square   Pulse

Sawtooth   Triangle

Figure 7.5

Of the four principal waveshapes, the sawtooth produces the richest amount of harmonics, and is, therefore, the most popular choice for creating many of the classic synthesizers sounds. From warm pads to gutsy synth basslines, the sawtooth wave can really excel, giving the sound plenty of

(Continued)

presence, depth, and bite! Be warned though, overuse of the sawtooth can make a track appear rather stoggy and one-dimensional, so it's worth exploring other variants and the sounds they can produce.

Compared with the sawtooth, a square wave produces a distinctive hollow and open sound to it, by the virtue of the fact that it only contains odd harmonics from the harmonic series (a sawtooth contains both odd and even). Acoustically speaking, this makes the square wave similar to the tone of an instrument like the clarinet; although it's also useful for bass sounds where it can sound "big" without grabbing large parts of the sound spectrum (and eating up your mix, therefore) like the sawtooth.

The pulse wave, a variation of the square wave, is created by changing the duty-cycle of a square wave, effectively producing a wave that both sounds and looks thinner and more nasal than the square wave. For example, the ES2's Oscillators 2 and 3 both feature a fully variable square wave, allowing you to morph from a conventional hollow square wave to the thinnest of pulse waves. One common tecwhnique is to modulate the wave's width using an LFO, producing an effect otherwise known as pulse-width modulation (PWM). Applied onto a pad sound, PWM can be a great way of adding a slash of warmth, somewhat reminiscent to the sound of chorusing.

A triangle wave creates the purest sound of the four types, containing just enough harmonic material to make it still worth putting through a filter. Triangle waves can be a great choice for a suboscillator, tuned an octave down from the main oscillator. Try this on a sawtooth bass, creating a woofer-shaking subsonic to the patch.

Amplifier and Effects (C)

The amplifier section – or in other words, the volume knob – controls the level of sound leaving the ES2 outputs. This section also includes a distortion unit (great for producing dirty, grungy, synth sounds) and a simple chorus/flanger/phaser unit for warming up any lifeless pad sounds (for example, this was a popular trick used on many of the older Roland synthesizers).

Modulation Matrix (D)

Left to their own devices, the Oscillators and Filter won't produce much interest to the sound by virtue of the fact that they are static – in other words, the sound doesn't change with time. By adding in modulation, however, we can start to add interest, dynamics, and movement into our patches. One simple example of modulation could be the process of linking keyboard velocity to the cutoff point of the filter – now, the harder you play, the brighter (or less filtered) the sound will appear. Other principal modulation sources include envelopes, which shape the ADSR characteristics of a sound over time, and LFOs, which can be used to add effects like vibrato or wah–wah.

As you'd expect, the ES2 features one of the most comprehensive and flexible modulation matrices going, with access to almost every facet of the ES2's programming architecture.

LFOs and Envelopes

The bottom section of the interface contains the controls for the three main modulation sources, including two LFOs and three envelope generators. The key concept to grasp here is that each LFO and envelope has a slightly different design, allowing each to excel in different tasks. LFO 2, for example, can be synched to the tempo of the song, while Envelope 1 provides quick and easy access to AD (Attack Decay) envelopes, which are particularly useful for percussive sounds.

7.4 Working with Oscillators

With an understanding of the overriding synthesis architecture of the ES2 under our belts, let's take a closer look at creating sounds on the ES2.

The first step in synthesizing your own sound on the ES2 should be to consider how you intend to deploy the various oscillators – do you, for example, just use a single oscillator, or use a number of oscillators richly detuned to create a thicker texture? Deactivating the oscillators can be quickly achieved by clicking on the numerical legend to the side of each waveshape pot, this will also have a corresponding reduction in the amount of CPU processing used, an important consideration for those using G4 processors.

As with many parts of the ES2, each oscillator can be seen as having some unique attributes, as well as features common to all three oscillators. Looking first at Oscillator 1, you should see a relatively standard set of synthesizer waveshapes, including sawtooth, pulse, square, triangle, and sine waves. One of the more intriguing options includes the wavetable at the bottom of the dial (preset to a sine wave), and a fully variable FM option (for more information on this, see Section 7.7). To audition the various waveshapes available as part of the wavetable, try clicking on the name and dragging the mouse up or down to scroll through the options.

Oscillators 2 and 3 add furthermore interest with the addition of a fully variable pulsewidth – morphing between a hollow square wave at the top of the dial and an increasingly thin and reedy pulse as you move the pot clockwise.

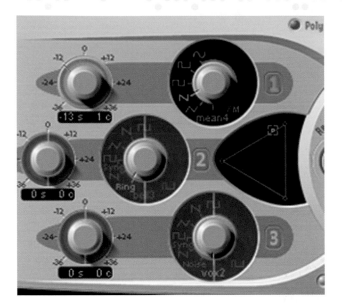

Figure 7.6
In this example, we've used just a single oscillator to create a simple synth bass sound.

Another important addition is the provision of two synchronizable oscillators, using either a sawtooth or square waveshape, controlled from the "master" oscillator – Oscillator 1. Going further still, Oscillator 2 also includes the option to ring modulate Oscillators 1 and 2, with Oscillator 2 carrying the ring-modulated output. Oscillators 3's "special feature" is the provision of a noise waveshape – a great source for the random harmonic information so important to percussion sounds.

In situations where you're using more than one oscillator, you'll want to experiment with their respective tunings. One simple technique is to apply a slight detune, say ±4 cents, using the fine control on the left of the tuning dials. This should fatten up the sound you're producing – the equivalent of three violinists, for example, playing the same musical motif. The use of semitone shifts adds furthermore interest and power to the patch. Try tuning an oscillator (set to triangle or square wave) down an octave (–12 semitones) from the main oscillators to create a subharmonic, or using third and fifth intervals (+3 or +4 for the third, +7 for the fifth) to create chord-like effects from a single note.

Beyond the basic detuning of the oscillators, there are a number of other features on the ES2 that can add width and warmth to the sound you're working with. The Analog control, for example, emulates the pseudo-random pitch and filter-cutoff drifts that occur within the imprecise workings of an analog synthesizer. Therefore, by increasing the Analog parameter, you can add an increasing amount of pitch and cutoff frequency drift, effectively making the patch appear warmer and less sterile.

Figure 7.7
Oscillator 1 includes a full wavetable made up of a number of digital waveshapes. Try Ctrl-clicking on the waveshape name to access the list as a menu.

Activating the Unison mode, in either of the ES2's monophonic modes (mono or legato), takes the "fatness" of your ES2 patches to a whole new level. As with the Unison control on a traditional analog synth, ES2's Unison stacks further virtual oscillators (on top of the three usually available in a patch) to create a thick, detuned, chorus-like effect. Going back to the string analogy, Unison is the equivalent of going from three violin players to a whole section, creating a sound that demands your attention in a mix. The actual width of the effect is defined in two parameters – the number of voices assigned to the unison, and the Analog parameter discussed earlier. In Unison mode, therefore, Analog

Knowledgebase 2 ▼

Oscillator Sync

Let's get this clear from the start: Oscillator Sync has nothing to do with synchronizing the oscillators to the tempo of the track, although of course, you can achieve this with the ES2's LFO 2. Truly, Oscillator Sync is "synchronizing" the wavecycles of Oscillator 2, with respect to the pitch Oscillator 1 – in other words, the pitch of Oscillator 2 is controlled by Oscillator 1. At first, this doesn't seem that exciting – simply controlling the pitch of one oscillator from another – however, by freeing Oscillator 2's tuning controls from conventional tuning applications, we open the lid on a whole new world of exciting effects.

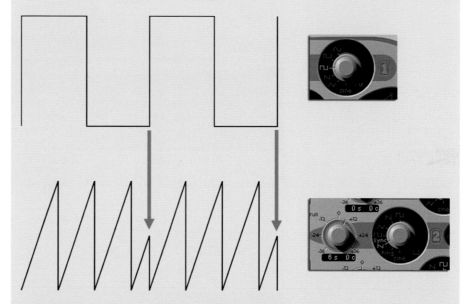

Figure 7.8

To fully understand Oscillator Sync, we clock what exactly happens to the pitch and timber of Oscillator 2 once it is placed in-sync with Oscillator 1. The effect is most clearly audible when the initial pitch of Oscillator 2 is set higher than Oscillator 1, so try initially tuning the Oscillator up 12 or more semi-tones. When the oscillator is placed in sync, you'll notice that its pitch return to that of Oscillator 1, although its timber has become brighter and harsher – almost distorted, in fact. Looked at under the microscope of an oscilloscope (try rendering the synths output as audio file and taking a closer peek in the Sample Editor), you should see the waveshape being forcibly retriggered, even though it begins its oscillation at a higher pitch. Therefore, it is this periodic retriggering that we perceive as the final pitch of the oscillator.

Oscillator Sync forms one of the most fascinating ways of distorting a waveshape, essentially creating a different version of the waveshape based on the relative pitch difference between the two oscillators. Although it can be used as a static effect, Oscillator Sync works best when the pitch of Oscillator 2 is modulated in some way (try using Envelope 1 for a classic "sync sweep" lead sound), creating dramatic timbral sweeps even before you've reached the filter!

changes its role somewhat and simply sets the amount of detune between the "doubling" oscillators.

Unison, of course, is a staple feature of many of the expansive lead synth sounds found in trance music, although used in moderation, it can also be great way of

Figure 7.9
Oscillators 2 and 3 include a number of interesting waveshape options – from fully variable pulsewidth to ring modulation.

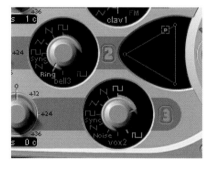

Figure 7.10 Activating the Unison control will layer several "virtual" oscillators on top of the three you're already been using in the patch. This creates a thicker and warmer output to the ES2.

creating some super-phat basslines. But, like all good things, Unison comes at a cost – namely, your CPU resources!

7.5 Filters, Amplifiers, and Modulators

With the initial pitch and timber defined by our oscillators, then comes the turn of the filters and amplifiers to shape the raw sound into something that grows and develops over time. The Filter section – found in the large central dial of the ES2 – removes unwanted harmonics from the rich, waveform-only output from the oscillators. For example, starting with a "buzzy" collection of sawtooth waveshapes, you could apply a strong low-pass filter to remove the high-end harmonics, leaving you with a soft and warm pad sound. Alternatively, use any one of the other filter types and filter-routing combinations – like band-pass, high-pass, or peak filter types – and you can achieve a real wealth of different timbers. Also, it's worth noting that the ES2 actually contains two independent filters – either allowing you to layer two different filter settings, or use two filters one after the other.

Knowledgebase 3 ▼

Filter Types

A large part of the "sound" we associate with a synthesizer is brought about by the performance of its filter. Filters come in many shapes and sizes, or more specifically, types and strengths, each with their own characteristics and applications.

The first thing to consider is the strength of the filter – which in the ES2's case is measured in three different decibel levels: 12, 24, and 36 dB. As you'd expect, the higher the decibel rating, the stronger the filter's ability to attenuate frequencies above or below the cutoff point. Interestingly though, a stronger filter isn't always the most desirable – most of the early Roland synths, for example, took a large part of their charm from the relatively "weak" 12 dB filters. A Moog, however, was famed for its steeper, 24 dB filter. On the whole, I tend to use the stronger filters on darker patches, whereas the classic, 12 dB filters work wonders on TB-303-inspired patches with lots of squelchy resonance.

The type of filter (Lo, Hi, Peak, band reject [BR], band pass [BP]) defines the form of frequency attenuation taking place. The principal filter type is arguably a low-pass filter (LPF), which attenuates frequencies above the frequency cutoff point – transforming the initially bright output of the oscillators into a more rounded overall tone. In most musical applications, the LPF makes most sense, and there are strong similarities to what happens within the acoustic instruments. For example, a piano note will start initially bright (filter open) with high-frequency energy decaying as the note decays (filter gradually closes). A high-pass filter (HPF), however, works in reverse so that frequencies below the cutoff point are attenuated. HPFs, therefore, are a great way of thinning sounds out, especially if the mix is starting to sound heavy and cluttered.

Band-pass and band-reject filters (labeled BP and BR on the ES2) are combination of LPFs and HPFs used to selectively modify a small band of frequencies. Band pass, for example, allows only a small band of frequencies to pass through, producing a characteristically small-sounding output. Band reject, however, leaves most of the audio bandwidth untouched, and instead attenuates just one "small" band of frequencies, much the same as a tightly Q'd dip on a parametric Equalizer. Conversely, a peaking filter applies a sharp boost at the given cutoff point, again similar to the EQ comparison, but this time a boost rather than a cut.

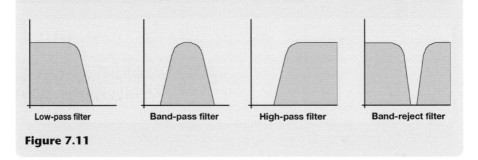

| Low-pass filter | Band-pass filter | High-pass filter | Band-reject filter |

Figure 7.11

Cutoff and Resonance

The two main parameters with both the filters are Cutoff and Resonance. Cutoff, as the name suggests, is the point where the filter begins attenuating frequencies – so, in the case of a low-pass filter, a lowering of the filter will move the cutoff point slowly down the audio spectrum, producing an increasingly darker output. In effect, therefore, you're letting the low frequencies "pass," while the higher ones are attenuated.

Resonance, however, creates a small boost around the point of the frequency cutoff, arguably making the filter's attenuation more pronounced. The exact effect of resonance, though, can vary both between the filter types and the relative position of the cutoff. Using the example of a low-pass filter, a high-cutoff

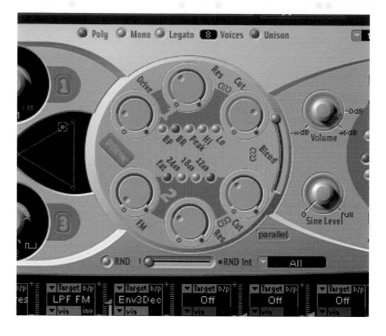

Figure 7.12 The filter section shapes the timber of the raw output from the ES2's oscillators section. The two main parameters to interact with are filter Cutoff and Resonance.

setting with some resonance applied will produce an increasing thin output (as the resonance boosts frequencies at the top end of the audio spectrum). Alternatively, resonance applied on a low-pass filter down the bottom of its scaling will produce an output with increasing amounts of low-end – an effect that can often be useful on bass sounds. Take the resonance all the way to its maximum setting, though, and the filter will self-oscillate (in other words, produce a high-pitched ringing sound almost like a sine wave) at a frequency defined by the filter-cutoff position.

Ultimately, the point to note with the filter is that it provides a large part of the "character" in a synth sound, probably best exemplified by the dripping, acidic resonance of a TB-303 filter in action on a bassline. All sounds, to a greater or lesser degree, will use the filter and if you can add some dynamic movement or expression into the patch (maybe responding to velocity, for example, using an envelope and/or LFO), this character will become even more pronounced and interesting to the ear.

Changing Volume Over Time

By default, the amplifier is controlled by Envelope 3, with the two remaining envelopes available for use with any other application you see fit. Envelopes allow you to manipulate the amplitude (or any other parameter, for that matter) over time – raising and lowering the given level for the duration for which the key is held down. The progression over time is defined by a number

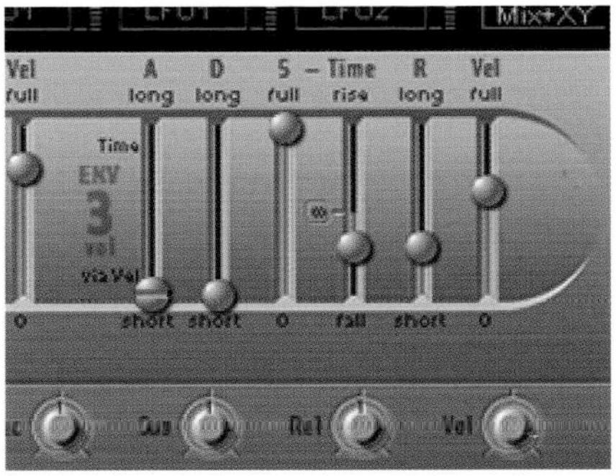

Figure 7.13 By default, Envelope 3 is patched to the amplifier, allowing you to shape the volume of the patch over time.

of key parameters: attack, decay, sustain, and release. Attack, like the majority of the other ADSR parameters, is a time measurement, so that a low setting produces a quick attack (useful for sounds that are more percussive in nature), whereas a high setting produces a graduated attack that, for example, might be applied to pads. Decay defines the time taken for the note to fall down to the sustain level, while release defines the final drop to silence after the key is released.

Looking at any sound in the natural world, it's easy to see how we can begin to define their qualities by the characteristics of an amplifier envelope. A piano, for example, would have a quick attack, slow decay, no sustain, and a relatively quick release. However, a smooth, evolving string sound might have a slow attack, slow release, medium-to-high sustain, and a slow release. So, coupled with the basic waveform selection and the attenuation of the filter, we can really start to establish the defining properties and uniqueness of the patch we're creating.

Of course, alongside Envelope 3, the ES2 also contains two other envelopes, but to understand how these are used we need to look at the modulation matrix.

Bringing It All Together: The Modulation Matrix

Tying all the elements together is the modulation matrix, which links a number of modulation sources in the ES2 (like the aforementioned Envelopes, LFOs, external MIDI controllers, and so on) to destinations within the synthesis architecture. The complexity and depth of the modulation matrix is what sets the ES2 apart even from many dedicated third-party synthesizers, and provides an almost unparalleled amount of flexibility and creative potential for synthesis. The only catch, however, is that certain elements of the ES2's control panel often need to be connected through the matrix before they have any effect on the sound!

Knowledgebase 4 ▼

Vector Synthesis

Vector synthesis, originally developed as part of the Prophet VS synthesizer, was first envisaged as a unique system for adding movement into a patch. In the case of the Prophet VS, it used a series of four oscillators, or sources, labeled A, B, C, and D, which could be morphed, using a vector joystick control. By storing this movement as part of a vector envelope, the patch could then recall this dynamic movement creating timbral movement impossible to achieve elsewhere.

By switching the modulation matrix from Router to Vector, you can activate the ES2's own vector-synthesis features. The vector controls themselves work with both the triangular oscillator vector mixer (using three oscillators, rather than the Prophet VS's four), but they can also make use of an additional vector controller in the form of the XY pad. By default, the XY pad has no assignment, but by using the X Target and Y Target parameters, you could, for example, assign the X- and Y-axis to Cutoff and Resonance – providing complete gestural control over two parameters on the one pad! To create a basic vector envelope, click on any of the vector nodes and position both the oscillator mixer and the assigned XY pad. Now, select the other two nodes in turn, ensuring each has a different position for the oscillator mixer and XY pad.

To hear the vector envelope in action you'll need to uncheck the Solo Point switch – you should now hear the ES2 morphing between its various states. To add further nodes into the envelope, simply Shift-click on the time line. You can also set the time between any of the nodes by click-dragging on

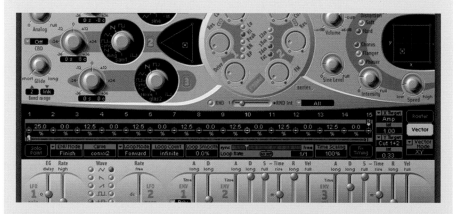

Figure 7.14 The ES2's vector envelope can add whole new levels of movement into your patch, by directly manipulating the oscillator mixer or XY pad.

the various ms settings, making certain transitions quicker than others. Changing the loop mode will then allow the vector envelope to loop, based on the direction set and the relative position of the two markers – a sustain marker (where the envelope would hold, if no loop were set) back to the loop marker, indicating the full duration of the loop.

By exploring some of its deeper features, you can also turn the vector envelope into a surprisingly effective source of rhythmic pads and textures. Try changing the Loop Rate from "as set" to a tempo division like 1/2. Now, the vector envelope appears to pulse in time with the track, a feature that could really be exploited with further nodes (up to 15 are allowed) and some interesting tuning and waveshape assignments of the respective oscillators. If you're in any doubt as to the effectiveness of this, try listening to some of the Vector Rhythm (Curve) presets as part of ES2 Legacy folder.

Looking more closely at the matrix, we can see a number of different routing paths (10 in total) indicated by the line of blue boxes. Each routing connection will have a target parameter (which could be a parameter like Filter Cutoff or Pan), a modulation source, and the amount of modulation that can be applied. Just a quick glance through the respective menus for both the target and source reveals the depth of the ES2, with over 40 program targets and over 20 possible modulation sources, with everything from standard modulation sources, like LFOs and envelopes, to random generators (RndNO1 and RndNO2) and external MIDI controllers.

Some of the principal connections you might want to explore involve controlling the filter from modulators like Envelopes 1 and 2, and the two LFOs. The LFO provides a "periodic" form of modulation – in other words, a repeating continual shape or pattern that modulates the parameter, rather than an ADSR-type modulation with fixed stages. Arguably the best example of this is vibrato – where an LFO would be routed to the Pitch123 target. However, LFOs can also be an excellent source of Tremolo-type effects (routed to Amp) or a wobbling wah–wah (routed to Cut 1 + 2).

Figure 7.15 The ES2 modulation matrix allows us to map up to 10 modulation sources (LFO, envelopes, and so on) to different destinations (Filter Cutoff, and so on) within the synthesizer's architecture.

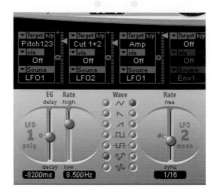

Figure 7.16 Each of the two LFOs can be set to a variety of different modulation shapes, each creating a unique form of modulation effect. For example, the sawtooth waveshape, or the stepped sample-and-hold waveshape can be great for sequencer-like effects, especially attached the filter. In that respect, it's worth noting that, in its below "DC" setting, LFO 2 can also be set to sync with the project's tempo.

7.6 Global Parameters and Output Effects

As part of the ES2 global parameters to the right of the filter dial we find a number of useful tools to add further layers of interest to the sound. The Sine Level, for example, adds an additional sine wave element to sound based on the pitch of Oscillator 1. This is a useful tool for general "bass fattening," especially if the output of your filter has left your patch sounding slightly weak. If you want more grit, though, the inbuilt distortion unit, which is switchable from hard to soft operation, helps in giving aggressive synth sounds considerably more edge. Even in relatively soft settings, with small amounts blend in, the distortion can make a patch sound slightly more analog than its digital origins would suggest!

The Modulation Effects section provides inbuilt chorus, flanging, and phasing effects, with a variable intensity and speed. All of these are useful warming effects, adding an extra sense of depth, stereo width, and movement to your ES2 patch that really seems to suit pad-like sounds.

7.7 EFM1 and Frequency Modulation Synthesis

Subtractive synthesis and the features of the ES2 might cover a large part of the sounds used in contemporary music production, but there are other methods of synthesis – like frequency modulation (FM) – with plenty to offer. EFM1 is Logic's solution for FM synthesis (although FM also makes some honorable appearance

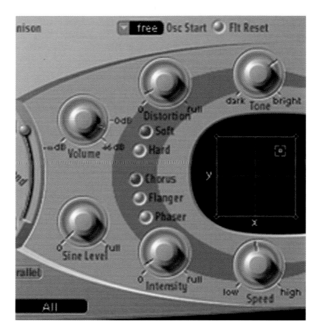

Figure 7.17 The output effects are a useful means of adding extra warmth (using the chorus unit) or grunge (using the distortion) to the ES2 patches.

in the ES2) using the same principles carved out in Yamaha's infamous DX range of synthesizers back in the 1980s. Although the EFM1 lacks some of the depth and complexity of a full-blown seven operator DX7, or Native Instruments' FM7, it remains a valuable insight into a unique form of sound creation and a valuable contrast to the sounds produced by the ES2 and Logic's other subtractive synthesizers.

Knowledgebase 5 ▼

ES2 Filter Routing

One of the most intriguing aspects of the ES2 is its immensely flexible filter-routing system. Essentially, the two filters can be connected either in series – so that sound flows sequentially from one to the other – or in parallel, where the sound is split and fed to each filter, individually. To switch between the two filter modes, simply click on the filter's Parallel or Series switch accordingly, and enjoy the sight of the revolving filter controls! Taking this up to another level, we then have the Blend parameter, alongside the Drive control, both adding a whole new level of interaction based on the current filter arrangement and its relative position.

(Continued)

Looking first at the Parallel routing (where sound is split and sent through to two filters individually), the Blend parameter can be seen as a simple crossfade parameter, making one filter's output more prominent that the other in the final mix. In serial mode, the concept is a little tricky to understand, although not impossible. In its central position, Blend creates a true serial configuration with Oscillator 1's output being passed to the Oscillator 2's. Either side of the line, however, an amount of either Filter 1 or 2 is progressively bypassed, effectively creating a bias toward one of the two filters. Of course, at either extremes of the Blend parameter, the output is exclusively Filter 1 or 2, respectively.

The Drive parameter adds polyphonic distortion either before or during the filter's operation (again, based on the filter's routing). Polyphonic drive differs from the conventional distortion (available in the amplifier section) in that it distorts each voice individually – in effect, using a different distortion module for each note played! In short, this means that a chord can be given a little "grit," rather than collapsing in intermodulate distortion, as it would in a conventional single distortion module setting. In Parallel routings, the drive is placed before the two filters, effectively, giving them slightly more harmonic material to play with. In Serial configurations, the Drive is generally placed after Filter 1 (although the exact interaction with filter 2 varies given the Blend parameter), making the distortion effect more pronounced, especially on high resonance settings.

Figure 7.18 Here are two ES2 synthesizers, each set to one of the two different filter routings – parallel or serial. Using the Blend and Drive parameters adds further possibilities to how you can use and abuse the ES2's filters.

Not surprisingly, the heart of FM synthesis, and the sounds produced by the EFM1, center on the process of FM – but what is it, and how can it be used to produce musically useful results? The process of FM requires two oscillators – one oscillator acting as a Carrier, and the other oscillator acting as a Modulator. You can see this clearly on the EFM1 front panel, with the carrier on the right-hand side of the panel, and the modulator to the left-hand side. The relationship between the two is that the modulator is modulating the pitch, or frequency, of the carrier, with the carrier oscillator being the only oscillator present in the audio output.

In many ways, the simple modulator/carrier configuration we see as a basis of the EFM1 bears a fundamental similarity to an LFO modulating the frequency of an oscillator on the ES2. As the LFO is working below the audio spectrum (around about 6 Hz), we hear the effect of this modulation as vibrato, with stronger amounts of modulation creating an increased amount of pitch wobble.

However, in the case of EFM, the modulator is working right up in the audio spectrum, producing an effect quite unlike vibrato. By modulating pitch using an oscillator in the audio spectrum, we effectively distort the original wave-shape, creating new harmonic components – the stronger the amount of modulation (set with the large central FM parameter), the greater the amount of the distortion and hence the richer the eventual harmonic output.

Figure 7.19
By increasing the amount of FM, we create a more harmonically rich output in much the same way as opening a filter on a conventional subtractive synth.

Figure 7.20 Changing the relative tuning of the modulator and carrier changes the resultant harmonics produced as the FM knob is increased.

One of the most important relationships to understand in FM synthesis is the ratio between the pitch of the carrier and the modulator. On the EFM1, the tuning of the modulator and carrier is established using the Harmonic control on either side of the interface. Try setting the carrier to the first harmonic and experiment with the modulator on first, second, and third harmonics. With both carrier and modulator set on the first harmonic, the sound will approximate a sawtooth, with a collection of both odd and even harmonics. Moving the modulator up to the second harmonic will produce a more hollow and open timber, with an output largely made up of odd harmonics. Going further still, this time up to the third harmonic, the sound becomes more bell-like, by the virtue of the fact that you're using an odd harmonic relationship.

Modulation on the EFM1

Okay, we've established the basics of a "static" FM sound – namely the tuning ratios between the carrier and modulator, and FM depth – but how do we go about adding movement or interest into the patch? Let's first consider trying to create a basic filter sweep. On an analog synthesizer, this would be achieved using an envelope patched to modulate filter cutoff, with the envelope opening or closing the filter over time. In the EFM1's case, we can use the Modulation Envelope toward the top of the interface in conjunction with the FM Depth parameter to the right of it. By setting up an appropriate envelope (quick attack, graduated decay, no sustain) with FM depth set to its maximum setting and the large FM dial set to its center position, the EFM1 should then gradually reduce the amount of FM as the key is held down.

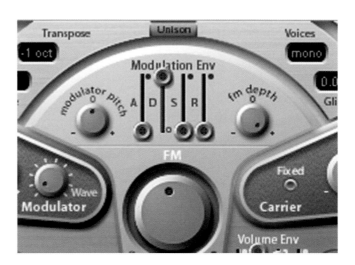

Figure 7.21 Use the Modulation Envelope and FM Depth parameter to modulate the large FM dial over time. The effect of this modulation would approximate a filter sweep on a traditional analog synthesizer.

Knowledgebase 6 ▼

Wavetable Synthesis

Wolfgang Palm originally developed the Wavetable synthesis in the early 1980s as a part of his groundbreaking work on the PPG wave synthesizer. The principle was that an "audio event" – like a filter opening – could be broken down and stored in a Wavetable constructed from 64 single-cycle waves. By stepping through these wave cycles, and interpolating between the various settings, the PPG could create unique and interesting textures, often characterized by a glassy, shimmering tone. Although PPG eventually went bust in the late 1980s, Palm's intriguing Wavetable system continued to live on (at least for a few more years!) courtesy of Waldorf's Microwave and Microwave II synthesizers.

By including its own digital Wavetable, the ES2 pays homage to this late great synthesizer, and, thanks to its flexible modulation matrix, can recreate many of the sounds that the PPG wave was so famous for. If you want to experiment with Wavetable synthesis, the trick is to select OscWaves as a modulation target and ensure the oscillators are each set to their Wavetable position. Choosing the exact type of modulation source (albeit an LFO, envelope, the modulation wheel, or something like keyboard velocity) will then dictate the type of movement and the sound produced. Try using a slow LFO (0.10 Hz, for example) and gently modulate the Wavetable portion to create a shimmering digital pad. Alternatively, you can create an interesting glitch-like effect by modulating the Wavetable using a fast AD envelope.

To add further interest into the equation, you can also modulate each of the Wavetables individually, possibly using a different modulation source (LFO 1, LFO 2, and modwheel) on each. To do this, you'll need to use OSC1Wave, OSC2Wave, and OSC3Wave destinations, respectively.

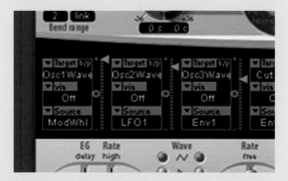

Figure 7.22 By modulating the Wavetables within the ES2, you can create some interesting "moving" synth sounds that are reminiscent of the PPG Wave 2.3.

Another form of modulation is the LFO, which, in the case of the EFM1, can be moved between being applied to pitch – to create vibrato – and being routed to the FM depth. As we've seen in the last example, routing a modulator to FM depth produces an effect comparable with filter modulation. In this case, therefore, routing the LFO to FM creates a form of wah–wah effect, with the timber of the EFM1's output wobbling over time. An entirely more chaotic form or modulation, though, is possible by routing the modulation envelope to the modulator pitch (using the dedicated pot). The effect is an extreme, and possibly dissonant, change in timber, which can be greatly used in aggressive synth sweeps.

Refining Your EFM1 Patches

To further refine your patch, the EFM1 includes a number of different parameters, which don't differ too greatly from the features found on the ES2. The Volume Envelope, for example, allows you to shape the amplitude over time – for example, try creating bell sounds using a quick attack, no sustain, and a slow decay and release. The suboscillator, as on the ES2, doubles the synthesizer's output with a suboscillator an octave below, giving bass sounds a real punch. Finally, the Stereo Detune option effectively creates a complete double of your patch, slightly detuned and with a distribution (of the two engines) across the stereo image. Used subtly, this can be a simple solution for adding warmth, depth, and width to any of your EFM1 patches.

7.8 Component Modeling: Sculpture

When it comes to cutting-edge sound design, Sculpture represents one of the most exciting and perplexing parts of the Logic, and also a real indicator of a future development of synthesis, and a component that may well become increasingly dominant in sound production as a whole. Sculpture's greatest strength – but equally the reason why it takes so much time to master – is that it throws out the conventional "rule book" of synthesis. You won't find traditional features like oscillators or amplifiers (although filters do get a minor input!), but instead parameters like Media Loss and Exciter Objects, producing sounds that are almost impossible to achieve elsewhere!

Like an increasing number of virtual instrument – including Arturia's Brass and Applied Acoustic Systems' String Studio – Sculpture is built on the concept of physical modeling, or, more specifically, component modeling. The basis of all these instruments is a complex mathematical model of an instrument, with a precise understanding of how sonic variables (the materials the instrument is constructed from, and so on) and various interactions (in other words, how it is played) affect the sound that the instrument produces. In the case of Arturia's brass, for example, the "model" is based on brass instruments like trumpet and trombone, whereas in the example of Sculpture, we have the multitude of possible musical outcomes that result from interactions with a string!

7.9 Objects

The starting point of patch in Sculpture is to consider how your instrument is constructed, and the means in which it will be played. Indeed, in the first part of designing your sound, you will spend a lot of time moving between two key areas of Sculpture's interface: the Material Pad, in the center on the control panel, and three different Exciter Objects, all interacting with each other and having some fundamental impact on the type of sound being produced.

Let's first look at the concept of Objects. To produce any note from a real instrument we need to introduce energy into it – for example, blowing across the reed of a clarinet or using a stick to hit a snare drum. Correspondingly, in Sculpture, we have a choice of up to three Objects that can be introduced to the string to begin creating a note: it could be hit, pluck, struck, or blown (to name but a few of our potential objects), with each different object producing remarkably different results. What this immediately means is that you need to have at least one object active to create any audio output whatsoever. Also, by having more than one object, you can create some weird-and-wonderful effects like a plucked string being gently dampened half way up its length, or complete audio chaos with a number of objects all playing their part at the same time! As you can see, the best results will come with a considered application of these settings, rather than just throwing it all "into the pot."

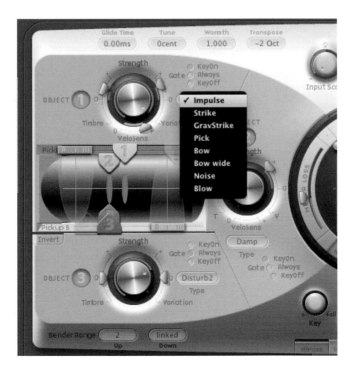

Figure 7.23
Sculpture's string can be played in a number of different ways, defined by the Type parameter of each object. For example, Object 1 includes Picking, Bowing, and a number of other "interactions."

Accompanying each exciter object, you'll find another three key parameters – Strength, Timber, and Variation – that set some of the initial effects of the object against the string. Of course, with each different object model the precise impact of the three parameters varies, so it is worth spending some time with each to understand the range of sounds and effects you can start to produce from them. If you've got the processing power, it might also be worth turning on the string animation by Ctrl-clicking on the thin green line toward the middle of the object controls.

Although you can produce some fantastic sounds with just the one object, the true potential and interest with Sculpture comes as you add further objects into the equation. In particular, you'll notice that Objects 2 and 3 contain some of the more esoteric object options, like Disturb 2-Sided, which approximates the effect of a ring placed around the string, or Mass, which places a large weight at some point along the string. Note also that you can set the object to be triggered in one of three ways – KeyOn, Always, and KeyOff – so that an object is only triggered after the key is released.

If you do use more than one object, it's worth paying close attention to the relative strength of the various objects. For example, using all three objects at maximum strength tend to create an overwhelming sound with little or no character. However, picking one key object (usually using Object 1) and augmenting this with subtle additions of other objects seems to produce the more

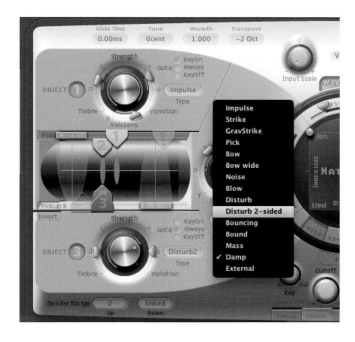

Figure 7.24
Objects 2 and 3 contain a number of different string interactions, which can be used to enhance the basic settings established with Object 1 – like the string being picked and damped at the same time.

interesting hybrid effects. For example, you could choose to use a Strike model on Object 1, which is then lightly Disturbed (the effect of placing an object on the string) using Object 2.

As important as the types of object used to initiate vibrations in the string, we also need to consider where along the string it is played. Take, for example, a nylon string guitar being played using a small pick. Played partway down the length of the string, just above the sound hole, results in a pleasing, rounded tone. However, if we were to play right up against the bridge, the tone will change dramatically, becoming harsher, with a thinner overall timber. Just as with real life, therefore, we can position each of the three objects along our virtual string using the Object Positioning display in the center of the Objects section.

Again, the best way of understanding how the models work, and their precise effect at each point along the string, is to experiment, although the general principle follows that a fuller sound emanates from the center of string, whereas the extremities tend to produce a more nasal timber. Adding in additional objects makes the variables ever more complex, although again, the general rule-of-thumb is that objects arranged closer together tend have a more pronounced effect on one another.

As part of the same object positioning display, you'll also find controls for each of Sculpture's virtual pickups. These work in much the same way as the electromagnetic pickups on an electric guitar, which of course, tend to output a different sound based on the relative position along the guitar string. In Sculpture's case, we're presented with two pickups that can be slid along the virtual string, contributing to the output in a number of different ways. To the far right of Sculpture's interface, you'll also find the stereo control, which can define a stereo spread both for Sculpture's key range (moving from left to tight as you move up the keyboard) and between the two pickups. Try using a wide stereo with the two pickups so as to create pleasing width to Sculpture's output – an effect especially useful on pad sounds.

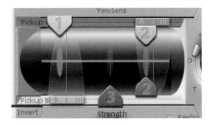

Figure 7.25 The position of an object can have a big effect on the string's vibrations. This interaction becomes even more complex with a greater amount of objects.

Figure 7.26 As with an electric guitar, the restive position of the pickups (A and B) will also influence the sound produced. Use the Spread control on the right-hand side of the interface to turn this into a stereo effect.

7.10 The String

Having spent so much time analyzing the objects used to excite Sculpture's string, we've neglected to look at an equally fundamental part of creating sound in Sculpture – the design of the string itself. Again, relating principles back into real life, we can see that the way a string is made will have a big effect on the sound produced, as evident in the apparent difference between a nylon-strung acoustic guitar and steel-strung acoustic guitar. By adjusting properties in the central Material Pad section, therefore, we can fine-tune the properties of our virtual string – morphing the sound between glassy, bell-like textures to warm, woody marimba tones, and plenty more besides!

The central X/Y controller of the Material pad sets the relative Inner Loss and Stiffness of the virtual string, although the precise qualities of this is better expressed by the small nylon, wood, steel, and glass legends in the four corners of the XY pad. Try moving the small control ball and listening to the harmonic spectra as you move around the pad – for example, listen out for the brighter tones toward the bottom of the XY pad against the move toward more bell-like harmonics as you move from left to right. As with objects placed along the string, this simple pad has one of the biggest impacts on the sounds you produce from Sculpture, so it's worth getting to know the various initial sounds possible from it.

Around the Material pad, we can also find a number of other key parameters in relation to the design of the string: Resolution, Media Loss, and Tension Mod. The most important of these is the Media Loss parameter, which sets the damp-ening applied to the string based on its surrounding conditions. A more mean-ingful way of understanding this would be the "release" of the sound, so that with a lower media loss setting, the sound will appear to decay over a long period of time, whereas quicker media loss settings tend to produce a shorter, more clipped sound.

Tension Mod corresponds to momentary fluctuations in pitch as the string is excited – the more the Tension Mod is applied, the greater these pitch fluctua-tions appear in the final output. Used subtly, Tension Mod can add a discrete sense of the natural pitch shifts that occur within a real acoustic instrument, although care should always be taken not to overuse features like Tension Mod as it could make the instrument sound like it has poor intonation! Resolution, however, corresponds to the amount of harmonics used to generate Sculpture's output. Try experimenting with lower resolution settings, as this can be an interesting way of creating inharmonic overtones (great for bell sounds) and as a means of reducing Sculpture's CPU load.

Figure 7.27
Using the central X/Y control will allow you to change the material properties of Sculpture's virtual string.

Figure 7.28
Parameters around the Material pad set a number of other physical considerations. Media Loss, for example, could be thought of as a form of decay or release setting, governing the loss of energy over time.

Taking the string design up another level, we can also add variations to its design based on the relative key position, otherwise known as Keyscaling, and in its release phase. So, for example, as you play up the keyboard the sound could change from a warm, woody timber to a bright, glassy string. Equally, Keyscale Media Loss could approximate the shortening of strings duration as you move up a piano keyboard, for example. To modify any of these settings, simply click on the Keyscale and Release bottom at the bottom of the Material Pad, and adjust the parameters (using the coloured indicators) as appropriate.

As a closing thought to the Object and String controls, try to remember that experimenting in component modeling synthesis is rather like relearning the art of synthesis. So, rather than using tried and tested means of solving problems – like adjusting an envelope to change the dynamics of a sound, or a filter to change the timber – you'll need to think "outside the box" and discover a new palette of solutions available in a component-modeling universe. Changing the amplitude envelope, for example, could involve a change or even an addition of an object onto the strings – a short decay, for example, made by physically dampening the string on the key off. Alternatively, adapting the string's Inner Loss properties, or simply less strength on the exciter, could make for a darker sound. Ultimately, it's all about new solutions to old problems.

Figure 7.29
Using the Keyscale and Release control, you can vary the properties of the string both in the release phase of the note, or in respect to keyboard position.

Knowledgebase 7 ▼

Bounce in Place

Rendering an audio instrument as an audio file has many advantages – not only saving valuable CPU resources, but also allowing you to safely export your project to other DAWs (or other users that don't share the same Audio Unit plug-ins), as well as opening up some interesting possibilities with respect to manipulating the resultant audio file.

Arguably the best way of rendering an audio instrument is to use the Bounce in Place feature, which renders one or more regions, complete with effect plug-ins, to a single audio file. To perform the bounce, simply select the desired regions in the arrangement area, and then choose Region > Bounce Region in Place. The accompanying dialog box will allow you the detail of how the bounce is made – including options to bypass and effect plug-ins, as well as the option to Normalize the file. Once you've performed the bounce, Logic will import the new audio file, place it accordingly, and assign it a new track. The source track can then either be left, muted, or deleted as you specify in the Bounce Regions in Place dialog box.

Once rendered as an audio file, you can now explore some of the possibilities unavailable in "virtual" form. Chief among these is the ability to reverse the region by entering the Sample Editor and selecting Functions > Reverse.

Figure 7.30 Use Bounce in Place feature to render an audio instrument as an audio file.

7.11 Waveshapers and Beyond

Theoretically, in a true component modeling synthesizer, the object and string parameters should be all that is required to create your eventual output. Indeed in Sculpture's case, it is highly likely that you'll spend the majority of your time working in these sections defining your sound, with the finished patch simply requiring a touch of EQ, compression, and reverb to set it in the mix.

Thoughtfully though, Sculpture also provides a number of additional features, including Waveshapers, Filters, ADSR envelopes, Delay lines, and EQs, to furthermore enhance and modify the effects you can produce within the plug-in. Although these aren't necessarily integral to the concept of component modeling, they are highly useful additions.

Following the Sculpture's internal signal path, we start with the Amplitude Envelope. The important thing to note here is that the amplitude envelope is post the component-modeling engine – its settings have no direct correlation to the sound produced and merely act as a means of modifying what's already there. So, for example, extending the release phase has no effect if the string has stopped vibrating (instead, consider raising the Media Loss), but a slow attack could be a suitable solution for making a bowed string have an even more graduated entry.

The Waveshaper, which follows after the Amplitude Envelope, is one of the most useful additions to Sculpture's synthesis engine, essentially applying a polyphonic distortion effect (much the same as the ES2's filter-distortion effect). In addition to applying a suitable valve-warming effect (try the Soft Saturation algorithm), the Waveshaper is essential in producing synth-like tones from Sculpture, making the output of the component-modeling engine much closer to the buzzy, harmonically rich output of a subtractive synth's oscillator. Of course, because the distortion is polyphonic (with each voice being distorted on an individual basis) even strong settings don't result in the sound collapsing in piles of unmusical intermodulate distortion.

Working in close conjunction with the Waveshaper is the filter, which includes five different filter modes – Hipass, Lowpass, Peak, Bandpass, and Notch. Given the added harmonic material created with the Waveshaper, you can now use the filter to subtract harmonics, producing synth bass sounds, and so on. Also the filter can be a useful additional way of tapering the output from

Figure 7.31
The amplifier envelope allows you to change the volume characteristic over time, but, remember, this is initially defined by the string and material properties.

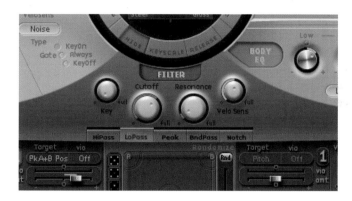

Figure 7.32 The Waveshaper adds grit or warmth to your Sculpture patch.

Figure 7.33 Use the filter as an additional form of timber control, with various different modes to change the relative effect achieved.

the component-modeling engine – maybe either using a LPF to create a dark bell patch or an HPF to take the low end out of a steel-string guitar patch. Remember, as with a conventional subtractive synth, the filter is also an ideal destination to control using the modulation wheel, an LFO, or an envelope, as we'll see in Section 7.12.

Three final modules provide the last refinements: a relatively simple Stereo Delay line effect (which isn't entirely dissimilar to Logic's Stereo Delay plug-in), a Body Equalizer, and a Limiter. The body equalizer's simplest application is Low Mid Hi Model, used to cut or boost frequencies at 80 Hz, 2.5 kHz, and 12 kHz, respectively. More advanced treatments are possible using one of the supplied instrument models, which are themselves drawn from impulse response recordings of the respective instrument bodies. In effect, what you have access to here is a detailed EQ profile of the characteristics of the instrument in question – it won't, for example, turn a harp sound into a cello simply through the application of the Cello's Impulse Response, but it does at least impart some of the instrument's tonal character.

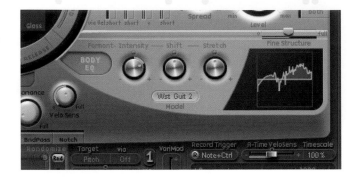

Figure 7.34
Use the Body
EQ and the
instrument-based
IR to impose
the timbral
characteristics of
an instrument
onto your
Sculpture patch.

Once loaded, the impulse response can be tweaked in a number of ways to best match the patch you're trying to create. Lowering the Intensity, for example, can be a useful way of softening the effect of the impulse response, imparting a subtle modification of tone rather than completely transforming it. Also, try experimenting with the Formant Shift and Formant Stretch parameters, allowing you to more carefully tune in the respective "lumps and bumps" of the impulse response profile with the tonal characteristics of what you're trying to create.

7.12 Modulation and Morphing

As we saw in the ES2, modulation is essential for bringing a patch to life: making it more expressive in response to actions on keyboard, adding in LFOs to add effects like vibrato and tremolo, and using envelopes to shape the sound over time. As you'd expect, Sculpture is no exception to this rule, with a modulation section (toward the bottom of the interface) that rivals, if not exceeds, what is possible on the ES2. More important, the application of modulation on Sculpture has the potential to directly influence the specific properties of the instrument (string construction, object position, and so on) rather than using crude "signal modifiers" (filters, amplifiers, and so on) to approximate the same effect.

The bottom dark gray pane in Sculpture contains all the various controls for changing the way the instrument responds to a performance and behaves over time: including two LFOs, a Jitter generator, vibrato, velocity modulations, controller routing, two envelopes, a Morph pad, and a Morph pad vector envelope – enough to keep anybody happy! Pressing the various tabs allows you to step through the relevant sections of the basic modulators, while the Morph pad and envelopes hide some intriguing "contextual" menus and functions, such as recordable envelopes and patch randomization.

Simple modulation can be accessed through the Vibrato and LFO 1 and 2 tabs. Vibrato permanently connects a spare LFO to the pitch of Sculpture, whereas LFO 1 and 2 can be wired to any number of destinations within the Sculpture patch – from the usual Filter Cutoff, and so on, to some of the more unusual

Figure 7.35
The modulation section of the Sculpture adds an additional level of dynamics and movement to your patch.

Figure 7.36
Routing the LFOs to the pickup position can be a great way of creating stereo movement in a patch.

component-modeling parameters like Object Position or Pickup positioning. Exploring modulation with these parameters, in particular, can lead to some of the most musically useful and rewarding applications of Sculpture, with complex or subtle modifications in timber that would be impossible to achieve on a traditional subtractive synthesizer. For example, try gently modulating the pickup position in a wide stereo pickup spread – the effect is a great way of adding subtle movement to a patch and palpable sense of "stereoness."

For direct response from a MIDI performance, Sculpture features two controller paths – labeled Ctrl A and Ctrl B – which are usually assigned to the Mod Wheel (1) and Foot Pedal (4), respectively, but can be changed to other controllers as appropriate. Again, think about creative and expressive uses of these routings – one favorite technique is to use two spare MIDI-controller faders to fully modulate the positioning of two objects on the string. This can be a responsive way of playing Sculpture, almost like a guitarist subtly moving a pick to various points around the sound hole and combining this with selective damping or muting on the strings.

The Jitter, Velocity, and Note on Random modulators can also be used to add variable amounts of chaos, unpredictability, and expression to your sound. Jitter, for example, can create small pseudo-random imperfections in the patch – try adding some Jitter routed to pitch to recreate the slight pitch variations exhibited on some acoustic instruments. Velocity and Note on Random can also be effectively mapped to parameters in the Material pad, so that different velocities, or alternate

Figure 7.37
Use the external MIDI controller to manipulate the relative positioning of the objects on the string. This can add a surprising level of expression to your patch.

Figure 7.38
Use Jitter to add pseudo-random modifications and drifts within your sound.

notes, produce subtle (or extreme) variations in the properties of the instrument. In theory, you could stack all of these modulation sources up, so that Sculpture responds and modulates in a multifaceted and complex way.

Although the modulation paths that we've discussed will start to bring some movement into your patch, probably, the best modulation features of Sculpture are to be found in the envelopes and Morph pad. The envelopes can be used in a conventional way (mapping to Filter Cutoff, for example), or alternatively, to record in modulation data – almost like an automation track within the architecture of the synth itself. To record an envelope into Sculpture, you'll need to activate the small R in the top right-hand corner of the envelope display. This places the envelope into record-ready mode, which, by default, can be initiated into recording by holding down a MIDI note and wiggling the modulation wheel, accordingly. Press R again to stop the recording, and you should see your moves stored into Sculpture's envelope generator.

In essence, the Morph pad stores a series of five snapshots representing positions of the main Sculpture parameters (Object Strength, Inner Loss, Tension Mod, and so on), which can then be morphed using the central XY pad. The great thing here, from a modulation perspective, is that over 20 crucial parameters can be adjusted at the same time (indicated by the small red dots). Now, add the fact that you can record these moves into an envelope to directly form part of the patch, and you have the world's ultimate vector synthesizer!

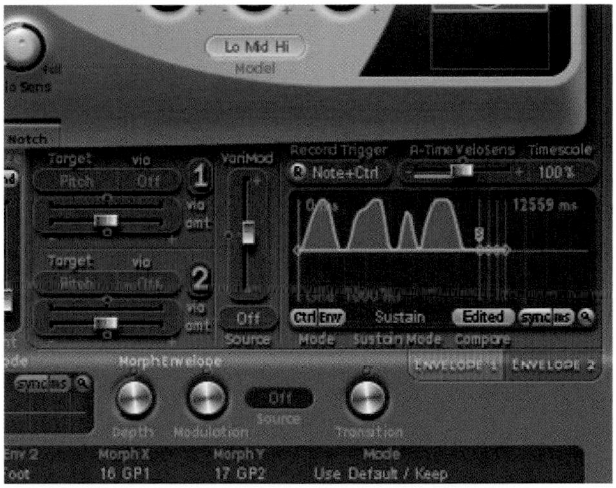

Figure 7.39
Placing the envelope into record mode will allow you to record your own "movements" into the envelope generator using the modulation wheel.

Arguably, the first step to working with the Morph pad is to define a number of different parameter positions for each of the Morph pad's corners (A, B, C, and D). Clicking on each respective corner, try to adjust the various key parameters, maybe applying differences in the Tension Mods, or the various positions of the objects on the string. The red dots provide reference to the current sound, also indicated by the position of the orange dot on the Morph pad itself. Note that you can also call up the Morph pad's contextual menu (by Ctrl-clicking), so as to copy and paste settings between the different points, or set the randomization destinations (initiated by the Rnd button) to have Sculpture "tweak" the various positions accordingly.

In some situations, you might find it adequate enough to have hands-on control with the Morph pad, and morph the patch as you see fit. Note that the MIDI-controller assignment section to the bottom of the interface details the standard MIDI assignment with Sculpture, allowing you to map the pad's XY axis to two physical MIDI controls on your keyboard. Moving beyond that, though, you might want to record your own vector envelope – unique to the Morph pad – that will then be permanently embedded into the sound. This can especially be effective on pads – with each note taking on its own "wandering" quality as the sound slowly shifts through different phases.

To engage the vector envelope, you'll need to change the Morph pad's mode from Pad to Pad and Env. As with the envelopes in Sculpture, you can record your own moves directly into the synthesizer. In this case, enable the vector envelope Record Trigger (using the small R), play a MIDI note, and wiggle the Morph point accordingly to record the moves in. Pressing R again will stop the recording, with the moves displayed on the pad to illustrate the movement between the various vector points.

Figure 7.40
Using the Morph pad you can define positions for each of Sculptures key parameters, based on your respective XY positioning.

Figure 7.41
Recording your own vector envelope will allow you to embed Morph pad movements directly into your patch.

7.13 Creative Sampling

Having explored some of the basics of using the EXS24 in Chapter 6, we now turn our attention to some of the more advanced creative aspects of using Logic's integral sampler. The assumption here is that you're using the EXS24 either to map samples (may be from a sample CD or recorded into Logic), or alternatively, you want to explore some of the deeper mapping and synthesis options within the existing EXS24 instrument library. Either way, you'll find the EXS24 to be a highly developed sound-design tool, and an interesting way of dynamically developing audio files that you've recorded into Logic.

The EXS24 Hierarchy: Zones, Groups, and Editors

As with any sampler, it's important to get to know the various hierarchical aspects of its operation – either in the form of different software pages or the way in which the sampler organizes and manages its sample data into zones, groups, and so on. One of the principal differentiations to make is between the main plug-in window – from which you can select instruments, apply envelopes, filters, and so on – and the EXS24 instrument editor window; which is specifically used for the mapping of samples across the keyboard. For most day-to-day activities, you'll only need to use the main plug-in window; however, if you intend to explore more creative applications of the EXS24 sampler, you'll need to make more extensive use of the EXS24 editor.

Within the editor itself, Logic also makes an important distinction between zones and groups. A zone is smallest piece of currency in the EXS24 editor, which essentially holds a selected sample mapped across a corresponding range of keys – like C3–D3. A group is a handy way of organizing the various layers of sample data that can be included in an instrument. For example, a group could be used

Figure 7.42
The two different sides of the EXS24 sampler – the main plug-in's interface window, and behind it, the EXS24 instrument editor where all the sample data is mapped.

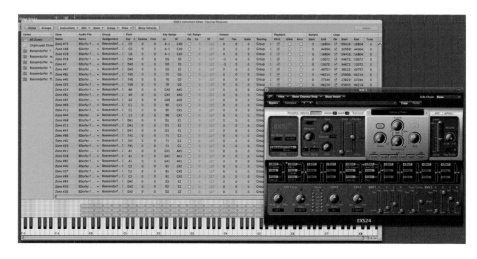

to differentiate between different velocity layers (soft and hard, for example), or two different components of a hybrid patch (like string samples in one group, and piano in another). There're also several interesting functional properties of groups – like the ability to apply different "conditional" triggering options, as well as the option to differentiate alternative filter settings, and so on.

Knowledgebase 8 ▼

Creating EXS Instruments from Audio Regions

As useful alternative to building EXS24 Sampler Instruments from scratch, consider using the Convert Regions to New Sample Track feature. This feature can work with a number of sounds – allowing you to transform a live drum performance, for example, or even vocals and acoustic guitar, using the power of the EXS24. To convert a region into a sampler track, you'll first need to select the region in question, and the pick Audio > Convert Regions

Figure 7.43 Use the menu option Audio > Convert Regions to New Sampler Track as a quick way of creating new EXS Instruments from an existing recording in your project.

to New Sampler Track. Logic will then give you the option to create a new sampler track based either on the region's transient markers, or the region as a whole. In most cases, you'll want to pick the transient marker option.

Of course, moving audio regions over to a sampler track is only the start of the creative process, especially given the variety of tricks and techniques possible between the EXS24 and the associated MIDI trigger data. The first option is to explore the various ways in which the source MIDI data can be manipulated and edited – including simple techniques like a different quantize values, or even going so far as to build a completely new "remixed" set of trigger points.

On the sampler itself, you can explore a number of different techniques using the EXS24's front panel – including adjustments to the filter, the sample's pitch, and the envelopes. Obviously, the filter can be used to provide a static filtering effect (making the loop darker with a low-pass filter, for example), but things get most interesting when some form of modulation is applied – whether it's note velocity modulating cutoff, ENV 1 shaping each note of the sequence, or one of the LFOs. As Envelope 2 is preassigned to volume, you can create some inserting effects by varying the Decay time, particularly with respect to short "choppy" envelope settings.

If your original audio region contains harmonic or melodic information contained within it, the relative merits of adjusting the EXS24's Tune parameter can be dubious. However, if your source material is percussive, or unpitched, the results can be far more interesting – transforming a clean drum loop, for example, into a crunchy collection of lo-fi beats.

7.14 The EXS24 Instrument Editor

The EXS24's sample editor can be opened using the small Edit tab in the top right-hand corner of the EXS24's main plug-in window – opening up the currently selected instrument in the editor. In many ways, it's a good idea to try opening an existing library instrument first – both as means of getting a grip on how the Instrument Editor presents its mapping, and also to see how Logic's sound designers have gone about using the zone and group features to create their instruments.

The EXS24 editor is divided into a series of areas, allowing you to best visualize and navigate the mapping that you've created. First of all is the two main view tabs in the top right-hand corner – Zones and Groups. These allow you to quickly move between viewing the entirety of the sample data and its mapping – zones, in other words – and the "macro-level" properties of a patch organized into groups. For more rudimentary EXS24 patches, it might be that you only need to

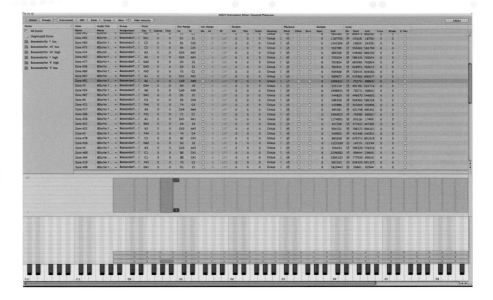

Figure 7.44 The Editor interface is organized into a series of areas, allowing you to visualize and edit the mapping of your samples in various ways.

work in the zones view. However, as your programming gets more complicated, it might become easier to manage by viewing the instrument in both the zone and group view.

In either view, the screen is divided into four key segments. At the top, you'll find the parameters area, which provides a long list of the samples or groups included within the instrument, alongside their associated parameters (like key range and so on) that define how they function. Moving down from this is the velocity area, which provides a quick snapshot of how samples are placed with respect to velocity – it might be, for example, that a zone or group works over the full MIDI velocity range (0–127) or only across a specific set of velocities (0–40, for example). Below this is the zone/group area that shows the placements with respect to the MIDI keyboard, which is viewable itself in the very bottom of the interface window.

Knowledgebase 9 ▼

ReCycle Files and the EXS24

Having seen how the Arrange area handles ReCycled files, it's also worth noting the distinct benefits and differences in working with REX files in the EXS24. Put simply, the reason for using the EXS24 over the Arrange page technique is creativity. As we'll see, once the REX files are imported into the

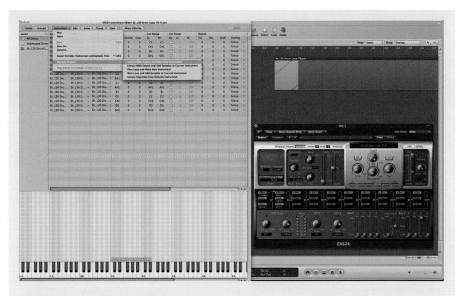

Figure 7.45 Importing REX files into the EXS24 is by far the most creative way of dealing with ReCycle files, facilitating a wealth of addition sound processing and resequencing options.

EXS24, you can start to manipulate the raw data in any number of ways, as well as take advantage of its tempo-tracking abilities by virtue of the loop being cut into a series of slices.

To import the REX file you'll need to be in the EXS24 editor, selecting the ReCycle convert option from the Instrument menu. Arguably the best solution to pick is the Extract MIDI Region and Make New Instrument option, which will not only make a new EXS24 instrument from the audio contained in the REX file, but also place a MIDI region file into your arrangement to trigger the samples in an appropriate way.

Once imported, the world is your oyster! Using the editor itself, you could decide to experiment with some of the zone parameters, with some of the most distinctive treatments coming from the selective reversing of certain slices. On the front panel of the EXS24, you could also decide to experiment with some of the synthesis features, maybe using the filter to make the loop darker, for example, or changing the envelope settings – maybe using a sharper decay setting with no sustain to produce a choppier output. Looking at the region itself, you could decide to resequence certain elements (removing or retriggering certain slices, for example) or even experimenting with different types of quantizing.

7.15 Creating a New Instrument and Importing Samples

Unless you're interested in remapping existing sample data, you'll need to create a new, blank instrument to work with. You can do this through the EXS24's Instrument Editor using the menu option Instrument > New.

Before you start to import sample data, you might want to consider some essential data management issues. For example, if you intend to keep your sample data in a unified place (like an external FireWire drive or a separate partition) you'll want to copy the sample data into an appropriate folder first, before initiating any imports into the EXS24. Of course, it is possible for the EXS24 to reference sample data from any drive – so if you're feeling lazy and have no intention of keeping your data in a unified place, just go on ahead and import! Finally, it's also possible to store sample data as part of your project's assets, which might at least facilitate a backup of the sample data, although there's no guarantee that the backup will exist on the drive 3 months down the line!

The importing process itself can be done in a two principal ways – either through the Zone menu (usually when you intend to load multiple samples in one action) or using a drag-and-drop from the project's Audio Bin and/ or the Browser. If you've invested the time in naming your samples properly, or indeed, chosen to import samples from a third-party collection, then the Zone > Load Multiple Samples is probably the quickest and easiest solution. From this option, you'll be able to specify the samples to be loaded, followed by a dialog to provide some indication as to how these should be mapped. The options are relatively self-descriptive, but see the following text for clarity:

Use Auto map for a collection of keyboard samples where the samples span more than one key. In this case, Logic will map the sample intelligently (using

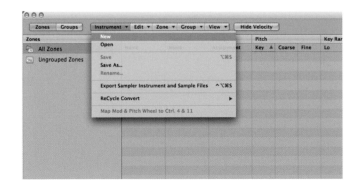

Figure 7.46 To build a sample mapping from scratch, select the New option from the Instrument menu.

the root key included in the samples' name) running each zone into (but not over) the next.

The Drums option again auto maps the samples based on the root key in the sample's name, although this will be the one and the only key the sample is mapped to.

The Contiguous zones is useful where you don't have key information included with the samples, or simply want to ignore any existing key information included with them. Using the Contiguous zones option you'll also need to specify the zone width for each zone (this could be one note, or a whole octave or more) as well as the start note to begin the mapping.

To build the instrument mapping up in a more gradual way, use the Browser window and drag-and-drop the files as you see fit. The exact way the samples are mapped will vary as to the samples selected and how you drag them into the EXS24 instrument editor. Dragging a single sample directly into the zone parameters area, for example, will place the sample across the entire keyboard with its default pitch at C3 (unless specified in the file name). Alternatively, drag a sample onto the MIDI keyboard on the bottom of the editor window, and you'll be able to map it to a specific key, although the range won't extend beyond that single note. Dragging onto an existing zone will also preserve the current zone parameters, but use the newer choice of sample.

Using the browser, you can also select multiple samples and import these directly into your mapping. As with the dialog from the Zone menu, you can also specify how Logic auto maps the samples. Note that by dragging the multiple samples onto a key on the MIDI keyboard, you'll effectively set the starting point, with subsequent zones following on from this point.

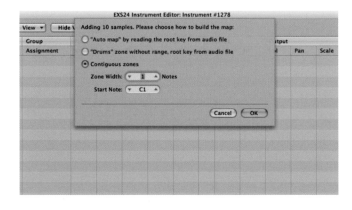

Figure 7.47
Using the mapping option might aid the quick placement of multiple samples – say from a sample CD.

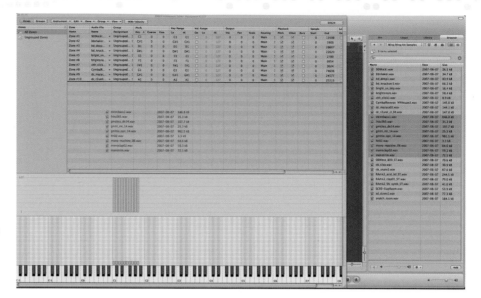

Figure 7.48
Besides using the Import option from the Zone menu, you can also drag and drop samples into your EXS24 instrument directly from the browser.

7.16 Changing Zone Properties

As part of the importing process, it's highly likely that Logic will have performed some basic zone parameter adjustments. Let's take a look at the different sections in the zone parameters areas and how you might go about using them.

Key Range

As we've already seen with the process of drag-and-drop sample assignment, each zone is assigned its own unique key range, with both upper and lower settings. If you're creating a drum-based instrument, it's likely that the Lo and Hi setting will be on the same note. However, if you're assembling a multisampled instrument – maybe with two or three samples for every "played" octave – you'll need to adjust the range over a few consecutive notes. You can adjust the Key range through the zone parameters area, or indeed, by dragging the appropriate "mapping blob" in the bottom zones/groups area.

Pitch

Pitch defines the fundamental root position of the note – in other words, the note on the keyboard that plays the sample at its natural sampled pitch. In the above example of a multisample spread over a few notes, you'll probably want to place the key somewhere in the middle of the zone's range, so as to keep the amount of transposition at a minimum. You can also use the Coarse and Fine tuning parameters to fine-tune the pitch of the sample – maybe the instrument

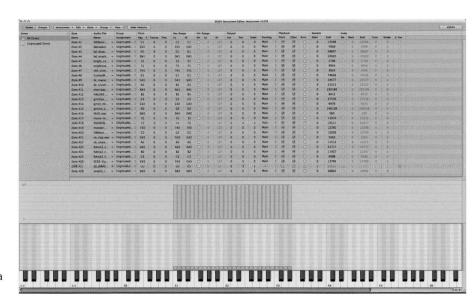

Figure 7.49
A drum-based
map will place
each drum hit on a
different note.

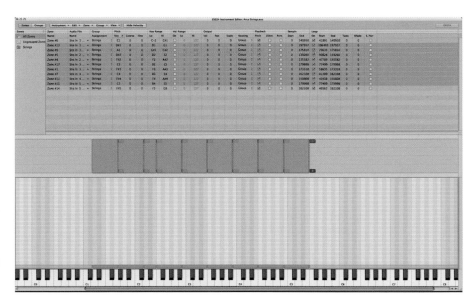

Figure 7.50
A keyboard-based
multisample might
use one sample to
cover two or three
notes – in this
case, the key range
should be adjusted
accordingly.

has been recorded slightly out of tune, for example, or you want to explore some more creative aspect of tuning on drum and percussion samples.

Vel. Range

Zones can be set to work over a specific range of velocities, using the Vel. Range parameter. If you've got several samples taken at different velocities, this might be a useful way of achieving a realistic sense of expression in your instrument, although you can always "fake" velocity sensitivity by some careful use of the amplifier and filter responding to velocity.

For more complicated patches – where you might be dealing with several different velocity layers – consider applying the velocity switching through the group feature.

Output

Output allows you to fine-tune some important mix-related issues with respect to your zone. Volume and Pan are largely self-explanatory, but what's of particular interest is the Routing option. By default, each zone is set to the main outputs, but it's also possible to force a zone out to any of the EXS24's 16 individual outputs. To access this, though, you will need to instantiate a Multi Output version of the EXS24 in your instrument channel, and activate the additional Aux channels accordingly.

By routing a zone out in this way, you'll be able to gain individual access to audio plug-ins as well a discrete channel fader for the zone(s) in question. This is particularly useful editing tool for drum-based instruments in the EXS24 library that you might want to enhance with additional processing, like compression, equalization, or reverb.

Playback

Playback governs some important aspects about how the sample data is read, mainly in relation to triggering drum samples. For example, the Pitch parameter allows you to enable or disable pitch tracking with a sample, so that even if a drum sample spans several keys, for example, its pitch remains fixed. 1Shot, however, focuses on the duration of the sample, so that in 1Shot mode the full length of the sample is played irrespective of how long the key is held down for. Again, the main application of 1Shot mode is with drums, where the duration of a hit isn't "variable" as such – in other words, a cymbal will always decay over a number of seconds.

Rvrs (or reverse) allows you to flip round the playback of the sample data. Given the need to create new audio files before you can reverse region in the Arrange area, this EXS24 reverse function is a quick-and-easy solution in creating reverse effects. Needless to say, the Rvrs feature is a great creative tool, sounding great on a range of instruments – from drums hits, to pianos and bells.

Sample

Although most samples that you'll import will have their edits already in place, it is possible to further refine the start and end point from within the EXS24 Instrument editor. For example, it might be the case that you have a small portion of silence at the front of the sample, or that you want to create a more abrupt end to the sample. Although the EXS24 Instrument editor displays this numerically, you can Ctrl-click on the sample number to open the sound file in Logic's Sample Editor. If you're running the editor at full-screen size, though, you'll move the EXS24 editor out the way to reveal the Sample Editor.

Loop

The Loop feature is primarily used as means of creating an infinite sustain on sampled instruments like strings, and so on. Obviously, if a sample was taken for the full-note duration, this could make an exhaustive drain on RAM allocation (although some third-party libraries do this), so samples are taken of about 5 s or so in length, with the last 3 s looped over until the notes end.

In the case of the EXS24, you can define a number of parameters in relation to loop, with the principal aim of creating a smooth, glitch-free loop that shouldn't be particularly audible to the ear. The beginning point is always defining the Start and End point of the loop – ideally picking a section of the sample that doesn't waver or drift too much, but instead, has a clearly defined sustain. Although you can define a loop point numerically, the easiest way to do this is to open up the Sample Editor (by Ctrl-clicking on either the start or end parameter number) and perform this visually. Make sure that you've got the Search Zero Crossings enabled (under the Edit menu), so that any start and end points snap to a zero crossing point.

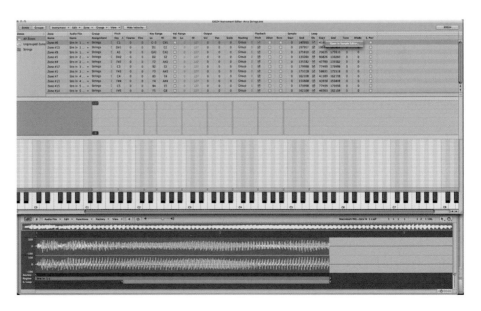

Figure 7.51
Although it is possible to set the loop numerically, opening the loop points in the sample editor (by Ctrl-clicking them) will allow you to better place the start and end points to match what is happening in the waveform.

Even if you do find a relatively successful location to position the start and end points, it might still be important to use some of the EXS24's additional loop parameters to fine-tune the looping to produce an even smoother looping effect. Xfade, for example, can be useful if you're still hearing small clicks or jumps around the loop point, with a larger number creating a longer crossfade. E. Pwr further refines the crossfade, so that no energy is lost in the crossfade processing, while tune addresses any tuning issues that might have become evident in the looping process.

7.17 Working with the EXS24's Groups

As soon as you start building up instrument files with more than 10 or so zones, it becomes essential to start to manage the samples in some meaningful way. Groups offer a variety of benefits to the EXS24 sample list – from organizing better ways of displaying related zones, to advanced triggering options. Using the instrument editor you can create groups in much the same way as you might create playlist in iTunes – simply select the zone or zones you want to group and then drag these across to the zone column on the left.

In a rudimentary way, the zone column offers an excellent way of navigating your instrument – either viewing the complete set of zones in an instrument, or choosing to view the instrument group-by-group by clicking on the appropriate

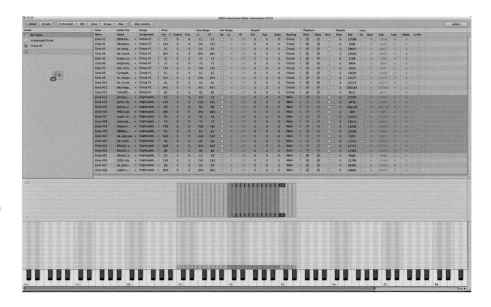

Figure 7.52
Select the zones you want to place in the group then drag them over into the zone column to create a corresponding group.

group folders in the zone column. Note that you can also name the groups by double-clicking on the group name. This is also an excellent way of organizing and creating hybrid EXS instrument – selecting different layers of zones from one instrument and copying and posting them into another.

Advanced Mapping Options

Moving over to the groups view in the EXS24 instrument editor, you should also see a number of parameter options unique to groups you have created. What groups offer is a form of macro-level control over an instrument – so, rather than having to individually set velocity switching over hundreds of zones, we can place them into four respective groups and then simply alter the velocity setting of those individual groups. Not only are grouped velocity adjustments far quicker in the short term, but also for any adjustments later on, the group feature is invaluable.

As you'd expect, the Key Range, Vel. Range, and Output parameters function in much the same way as they did in the zone parameters, although of course, settings here override above what you've established in the zone (for example, if the group's key range doesn't extend as far as where you placed a zone, you won't hear the sample).

The Trigger option defines whether a group is triggered by a Note On or a Note Off command (known as Key Down and Key Release, respectively) – an effect more broadly referred to as Release triggering, which uses short samples of events that occur after the main note to create a more real-istic and life-like "musical event." Release samples could be in the form of reverb, for example, or the small decaying "resonance" when a piano note is released.

Following on from the trigger options – although positioned over the far right-hand side of the group parameter – are the conditional Select Group By options. Although you can create a number of different conditional triggering options here, probably the most well known is the option to create a so-called key-switched instrument. With key switching, a number of notes down the bottom of the keyboard will be used to dynamically toggle between different instrument

Figure 7.53 Groups contain a unique series of mapping options – including filter and envelope offsets and various conditional triggering options – ideal for the expert EXS24 user.

articulations (like staccato and legato), all on the fly. Each articulation, as such, is organized into its own group, with the group "selected by" a specific note, like C1.

The Select Group By actually works with a number of different MIDI events (Note, Controller, Pitch Bend, and Channel). This makes conditional triggering an incredibly flexible feature – allowing you to use a modulation wheel, for example, to switch between layers.

The Envelope and Filter Offsets allow you to differentiate between the various groups and the initial settings on the EXS24's front panel (which we'll look at in more detail later). For example, although you might set a master filter cut-off position at 50%, for example, you could use the Filter Offsets parameter so that one group opens the filter in a more open position (+50), while the other closes in down (−25). Although this isn't quite the provision of a completely independent multimode filter per group, it does allow different zones to use different filter settings concurrently – in other words, while one group sounds filtered, the other is less so.

Following this concept further are the envelope offsets, which again work from the basic premise of the parameters established on the main front panel, but then adapted, negatively or positively, to match the specific needs of the group. You could, for example, have a nylon-string guitar in one group using the fast attack and release settings, while in another group, some string samples use softer attack and release settings to produce a lush background pad.

7.18 Editing EXS24 Instruments Using the Front Panel

With the samples mapped, now comes the task of adjusting and refining the synthesis engine of the EXS24's sample playback. The settings on the front panel extend what you've already achieved in the mapping of the samples. For a simple instrument, for example, all you might need to adjust is a few key parameters like release, or its velocity scaling. Alternatively, you could use the front panel of the EXS24 to completely transform the samples way beyond their original sound – maybe using some extreme filter and drive settings, combined with a series of creative modulation-matrix settings.

Like the ES2, the front panel is divided into a number of key sections, including the filter block, the modulation matrix and the LFO and envelope modulators. To understand how these parameters work, let's look at a couple of different editing scenarios and how these might work.

Basic Instrument Refinement: Velocity Scaling, Polyphony, and Envelopes

Not all instruments require a great deal of modification away from the default settings that appear on the EXS24's front panel, but there are a couple of areas worth looking at to best optimize the instrument in the final mix. First of these has to be the volume controls over the right-hand side of the plug-in interface. Alongside the overall volume level (which might be worth reducing if you're using a lot of sample voices simultaneously), you'll find the sampler's scaling of level through velocity. Raising the bottom part of the level through velocity slider can sometimes be a useful alternative to compression, especially if you've got a multisampled instrument with "quieter" samples in its low velocities.

Adjusting Envelope 2 – which, by default, is assigned to the Amplifier duties – is a useful way of quickly grabbing hold of, and shaping, an instrument's ADSR characteristics. For example, if some of the samples in your instrument have reverb embedded into them, you might find it beneficial to increase the release time so that you hear the full decay of the room ambience. Alternatively, using a slower attack can be a valid sound design treatment to soften or disguise the initial attack transient of a sample.

Toward the top left-hand corner of the interface are a number of useful macro controls, some of which can have a big impact over the performance of the sampler. For example, the voices parameter reserves a given amount of polyphony for the instrument in question. Sixteen notes, for example, might be more than enough for a simple drum-based instrument, but should you use something like a multisampled piano with release samples, you might need to raise this up to the full 64 voices. However, with an increased voice count comes an increased CPU drain, so it's often the case that you actively try to balance out the realistic demands of the instrument, against the overall needs and drain of your CPU performance. A used indicator at least provides some rough indication as to the total used polyphony, but you should also be able to hear a "maxed-out" polyphony by longer notes appearing "snatched off."

Adding in the Filter

As we saw with the ES2, the filter has a surprisingly powerful way of shaping the timber of our sound source. The filter included in the EXS24 is certainly no exception to this rule: with up to six different modes of operation (including high-pass, band-pass, and four different strengths of low-pass), some fat and juicy resonance, and even a Drive parameter.

Figure 7.54
The main controls in relation to volume are on the right-hand side of the EXS24's interface. Try experimenting with the sampler's response to velocity, as well as how Envelope 2 changes the amplitude over time.

To activate the filter, you'll need to press the On switch, otherwise it defaults to its off position so as to save a few CPU cycles. Switching the modes can be done using the six tabs at the bottom of the filter, with the additional Fat button to preserve some of the low-end with high-resonance settings on the LPFs. Remember that the precise position of the filter cutoff and resonance can also be varied between different groups, using this front panel setting as the initial starting point.

Modulation Heaven!

Many of the more advanced synthesis options come through the application of the modulation matrix, which allow you to map a variety of sources (including the LFOs and envelopes at the bottom of the interface) to various destinations within the EXS24's synthesis and sampling engine. As with the ES2, the EXS24 uses a familiar matrix display, with a Destination parameter (like Filter Cutoff), and amount of modulation and the Modulation Source including any of the

Figure 7.55 Once you've engaged the filter, you can use its variety of modes (low-pass, band-pass, and high-pass) to shape the timber of the EXS24's output.

Figure 7.56 As with the ES2, a large Modulation Matrix section allows you to route a number of different modulations sources to targets in the EXS24's programming architecture.

EXS24 modulators (three LFOs and two envelope generators) or any one of the 127 standard MIDI controllers.

Saving Instrument Settings

Confusingly, the EXS24 makes some level of distinction between the instrument settings stored with and without the front panel controls. To permanently store the front panel setting alongside your instrument mapping, you'll need to go to the Options menu on the EXS24 and select the option "Save Settings to Instrument." Now, when you return to that instrument,

Figure 7.57 If the settings on the EXS24's front panel have become intrinsic to the instrument in question, consider using the Save Settings to Instrument feature under the Options menu.

the mapping will be loaded alongside any front panel settings that you've created. Alternatively, use the Recall Settings from Instrument or Recall Default EXS24 Settings if you've made changes to the front panel settings that you're not happy with.

7.19 Ultrabeat

Ultrabeat is Logic's take on a contemporary drum machine, complete with a range of synthesis and sequencing options, as well as a handy multichannel version to allow the flexible access to Logic's signal-processing features. An Ultrabeat Kit is made up of 25 different drum voices – listed down the left-hand side of the interface complete with level controls (as the blue bars), mute, solo, pan, and an output selector. On selecting a drum voice, you can then edit its associated synthesis parameters in the large central display, or indeed, program its unique step sequence at the bottom of the plug-in.

In keeping with the flexibility of the ES2 and Sculpture, the synthesis options in Ultrabeat are impressive to say the least! Each drum voice has access to up to three different oscillators, which are then passed through a multimode filter a ring modulator, and finally some EQ. Arguably, the most useful oscillator, though, is Oscillator 2, which features a sample playback mode as well as the phase Osc and model modes. On the whole, you'll find that samples offer the most quick and effective solution for assembling a kit (if so, use the "blank" Drag & Drop Samples Kit from Ultrabeat's plug-in settings), although it's also good to augment these with some of the pure synthesis elements, like the noise generator. Once the sample mode is active, you can also drag sample files directly from the browser onto Oscillator 2.

Figure 7.58
Ultrabeat features up to 25 different drum voices in each kit, each with their own set of synthesis options.

Figure 7.59
Use the Browser to drag and drop samples directly into Ultrabeat's second oscillator.

Clicking on the Filter tab activates the filter, offering a great way of coloring the samples, like a band-pass filter applied across a clap sample to make it sound narrower. To modulate the filter, you can also use any one of the four envelopes and an LFO, with the small blue legend beneath the cutoff and resonance controls, allowing you to define an appropriate modulation source. If you're after grit, though, try activating either the crush or distort modes toward the bottom of the filter dial – both effects that can sound great on heavier kick or snare samples that you want to push to the front of the mix.

When it comes to assembling the sequence, you can just step through each voice and create the finished pattern piece-by-piece. A much more effective solution, though, is to switch Ultrabeat into its Full View mode using the small tab in the bottom left-hand corner of the interface. In full mode, the synthesis controls are removed to reveal a full-sized drum grid for all of Ultrabeat's 25 drum voices. This makes it far easier to gauge the interaction between the various parts, rather than having to remember their various hit points. Note that to hear the sequencer in action, you'll need to activate its Power tab, with the transport controlled either from Ultrabeat itself (in which case it plays independent of the rest of the project) or from Logic's main transport, so as to audition the pattern in the context of the song.

Figure 7.60
Use Ultrabeat's Full View to sequence the complete pattern, which can then be dragged directly into Logic's Arrange area.

In total, you can sequence up to 24 different patterns with Ultrabeat, with the potential to switch between them using a MIDI keyboard. A much better solution, though, and one that makes more sense in the Arrange area, is to export the pattern out as a region. To do this, click-and-drag on the small icon next to the pattern selector, placing the resultant region on Ultrabeat track in the Arrange area. Exporting a number of patterns in this way, you can then revert back to Arrange area to structure your song accordingly, even using other MIDI editors (like Piano Roll or Hyper Editor) to further manipulate the data

In This Chapter

8.1 Introduction   259

8.2 Channel Strips: Understanding Your Virtual Console   260

8.3 Organizing Your Mixer: What You Do and Don't See   268

8.4 Folders and the Mixer   273

8.5 Beginning a Mix   275

8.6 Adding Send Effects   280

8.7 Combined Processing Using Aux Channels   283

8.8 Using Groups   285

8.9 Working with Channel Strip Settings   288

8.10 Automation: The Basics   289

8.11 Track-Based versus Region-Based   290

8.12 Automation Modes   292

8.13 Viewing and Editing Automation   295

8.14 The Automation Menu Options   300

Plug-In Boxouts

Distortion Effects   261

Helpers ... Handy Little Plug-Ins   274

Compressor   277

Channel EQ   280

Convolution Reverb and Space Designer   284

PlatinumVerb   291

Using Delay   294

Vocal Processing   297

Pedalboard   306

Plug-In Focus

To Gate or Not to Gate ... That Is the Question   266

Knowledgebases

Clipping Faders: Good or Bad?   270

Parallel Compression   287

Saving CPU: Freezing and More ...   299

Compression Circuit Types   302

Walkthrough

Adding Compression   303

Logic Tips

Plug-In Delay Compensation   305

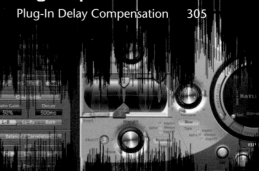

Mixing in Logic

8.1 Introduction

Up until a few years ago, there was only one accepted way of creating a professional sounding mix: hiring a commercial studio facility and mixing through a traditional mixing console, like an SSL or a Neve. However, times have changed significantly since then, and mixing "in the box" – in other words, mixing completely within the domain of your audio sequencer – has become a viable, and some would say more flexible, alternative to the traditional console route. In truth, applications such as Logic offer a tremendous amount of creative and technical freedom to today's mix engineer: complete and instantaneous recall, compression and equalization on every channel, studio-grade reverb, full automation, and much more besides.

In this chapter, we're going to take a closer look at the process of mixing in Logic, looking at both the technology it has to offer to create an effective mix and how we can knit these elements together as part of the mixing workflow. This is also an opportunity to get to know many of Logic's essential signal processing plug-ins in a more informed way, as well as taking a more detailed look at how the Mixer area works – aspects that can actually improve the entirety of your workflow in Logic, as much as the mix. Looking more closely at the process of mixing itself, we'll explore how to use compression and equalization to create separation, as well a bus processing (on aux channels), and of course, spatial treatments like reverb and delay that help define the depth and dimension of your mix.

Beyond the essential signal processing tools of mixing, we'll also take a look at features like automation, grouping, and folders, all of which help make your mix easier to manage, especially with the large track counts that most projects seem to encompass.

DOI: 10.1016/B978-0-240-52193-0.00008-3

8.2 Channel Strips: Understanding Your Virtual Console

Before diving headlong into your first mix on Logic, it's worth taking some time to distinguish between the various channel strips that the Mixer comprised and how these various elements interact with each other. Generally, a good mix in Logic makes use of all the features of the Mixer – channel inserts, bus sends to aux channel strips, output masters, and so on – so it's worth familiarizing yourself with their features in much the same way as an engineer would "get to know" the sections of a physical console.

Audio Channel Strip

An audio channel, as we've already seen in previous chapters, governs the basic path in and out of Logic for a recorded signal. In a mix, audio regions in the corresponding track will be sent down the channel strip, through the various insert processors (compression and EQ, for example) and bus sends (for reverb and other effects) to a designated output.

Instrument Channel Strip

The instrument channel strip duplicates the same features as an audio channel strip; only this time you get to work with the signals generated by the virtual

Figure 8.1 An audio channel strip.

Figure 8.2 An instrument channel strip.

instruments in your session. In addition to this simple application, an instrument channel strip can also be configured for Multi Channel operation so that multiple outputs from the same instrument (like the EXS24 or Ultrabeat) are sent to a number of additional auxiliary output channels. Besides having individual level control, this also allows you to apply different effects onto each output so that a snare, for example, might have different equalization, compression, and reverb to that of a kick drum.

Plug-In Boxout 1 ▽

Distortion Effects

Logic includes an impressive set of effects dedicated to low-fi transformations and crunching-up sounds – from the digital extremes of Bitcrusher, to the warmer tones possible with Guitar Amp Pro to the detailed emulation of many favored tones emulated within Amp Designer. Besides being an obvious addition to guitar parts, distortion can also be a great source of

(Continued)

color and interest on any number of sounds on a mix. Try using distortion to add some grit and body to a drum loop, for example, or a touch of drive on an aggressive vocal.

The first choice of amp-like distortion would have to be Amp Designer, which provides immense flexibility with respects to the choice of emulated head, speaker cabinet, microphone type, and its position. Additionally, there are a considerable wealth of presets to get you stared. Add to this the power offered by the new Pedalboard plug-in offering the guitarist endless scope.

Figures 8.3 and 8.4 Logic's Amp Designer offers an impressive selection of amplifiers to choose from.

Figure 8.5 Amp Designer with Pedalboard brings a whole new scope for guitarists in Logic 9.

Guitar Amp Pro, however, is still a solid option, which emulates the Amp, EQ, and speaker components of a typical guitarist's amplification system, as well as modeling a number of different mic'ing options. The clear advantage here, as with Amp Designer, is that the tone of the guitar can be completely modified (changing the Amp head, for example, or the position of the virtual microphones) at any point in the production process, without having to rerecord the guitar.

Distortion, overdrive, and clip distortion come from earlier incarnations of the application and tend to be less suitable for applications on electric guitar. They are, however, an excellent addition to drum loops or acidic synth lines generated from the ES2. Both distortion and overdrive have particularly simple controls with just a drive and tone parameter, alongside a corresponding output reduction slider to avoid ripping your monitors to shreds!

The prize for the most bizarre distortion effects goes to the Phase distortion and Bitcrusher plug-ins. Phase distortion sounds great on drums loops needing to be mashed up beyond all recognition. Based on a modulated-delay line, it can produce tones that range from a warm fuzz to the sound of a FM broadcast gone considerably wrong! Bitcrusher, on the other hand, is a great source of digital grit – especially as you reduce down the bit slider (to around 8 bits) and start applying some heavy downsampling (103 and beyond).

(Continued)

Figure 8.6 Distortion can bring plenty of color and grit to your mix. Try experimenting with Logic's wide range of distortion effects to achieve anything from a subtle valve-like warmth through to early digital distortion. Guitar Amp Pro still provides solid guitar tones in addition to Amp Designer. Logic's Amp Designer offers an impressive selection of amplifiers to choose from.

Aux Channel Strip

The humble aux channel strip is one of the most useful parts of Logic's Mixer and the source of a number of different techniques in mixing. First, it can be used as means of applying send effects like reverb or delay – anything where you want to actively control the balance of wet and dry sound using a separate fader. In this example, the aux channel strip has the effect strapped across its insert path with instrument channel strips or audio channel strips sending signals via designated bus sends. This allows any number of channels to access the same reverb, for example (an essential way of preserving DSP resources), as well as being able to control the return of the effect in the same way as any other audio signal in the mix – in other words, it could be equalized, compressed, and controlled by movements of the fader.

Another less immediate, but altogether just as useful, technique is to use an aux channel strip to combine a number of different channels (including audio and instrument channels) on a single fader. For example, many engineers using a traditional console will create a submix of the drums to a selected bus master fader

Figure 8.7 An aux channel strip.

(in this case, one of Logic's aux channel strips), allowing them to quickly control the level of the drums relative to other instruments in the mix. Additionally, they could also make use of inserts on the bus faders to apply compression, EQ, and so on to the entirety of the drums, rather than individual channels within it – something which can just as easily be done of Logic's aux channel strips.

The final important role of the aux channel strip is to operate as a means of inputting external signals into the mix. If your audio interface supports enough inputs, there's no reason why you can't patch-in external compressors, equalizers, synthesizers, and so on directly into your mix. Although you might need to account for a small amount of latency (as discussed in Chapter 3), using dedicated recording hardware can often supply much more character than conventional plug-ins. Of course, given the provision of inserts on these aux inputs, there's no reason why you can't also use Logic plug-ins on top of whatever device you're inputting into that particular aux input.

Stereo Output Channel Strip

The stereo output channel strip represents each of the physical outputs in your Logic system. If you're using a simple two-in, two-out USB audio interface, you'll only see one of these faders. However, if you're working with a multiple output FireWire or PCIe soundcard, you can increase the number of output faders corresponding to each physical output. On the whole, most mixes simply use one designated output (1–2) as the "destination" of the two-track mix. If you intend to render the mix directly from the Logic session, you'll also need to make use of Bounce button, found on the output channel strip or File > Bounce or simply Cmd + B (for more information on bouncing to disk, see Chapter 9).

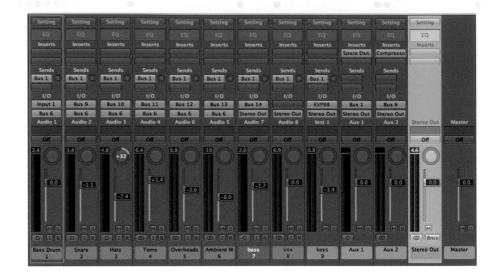

Figure 8.8 An output channel strip.

Plug-In Focus ▼

To Gate or Not to Gate … That Is the Question

Music trends change, so it's not a surprise that the process of recording also adapts concurrently. For example, just a few years ago, it was commonplace to gate the components of a drum mix (kick, snare, toms, and so on) to within an inch of their existence! Nowadays, it seems that listeners, engineers, and musicians all seem to prefer a looser sounding kit, resulting in much less use of gating.

Now and again, though, you may well feel the need to improve your aural hygiene – maybe a vocal microphone has some obtrusive background noise made worse through the application of compression or that one and only guitar take has a little too much noise from the amp. If you're pushed for time, there's no doubt that the noise gate plug-in remains a quick and effective way to attenuate problematic noise in between notes – although remember to gate before the compressor and not after it! The trick with setting the noise gate is to start with hard settings (quick Attack, Hold and Release, with 100 dB of reduction) to find the right threshold. Although the gating sounds harsh, you'll be better able to find the "sweet spot" just above the amplitude of the noise. With this set, back off some of the settings, especially Release and Reduction to soften the effect – surprisingly, even a small amount of reduction (6–10 dB) can have a big effect on the overall cleanliness.

If you've got a little more time, most users now tend to approach the gating issue through a few crafty edits. Try using the Arrange window's Strip

Silence option (Audio > Strip Silence) as a means of Logic preparing the majority of edits for you. As with a conventional noise gate, you'll need to establish the right threshold to get the best results, although arguably this is easier to "see" on Strip Silence than it is to "hear" on the noise gate. However, the big advantage with the Strip Silence approach is that you can modify the edits later on, making sure none of the important attack transients are missed out.

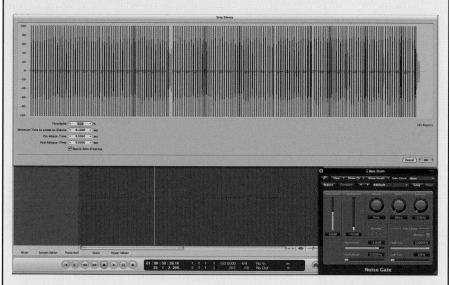

Figure 8.9 Two choices for improved aural hygiene – the traditional noise gate or the current favorite, Strip Silence.

MIDI Channel Strips

Arguably, MIDI channel strips should be kept distinct and separate from the audio mixing, but given their inclusion as part of the Mixer, it's worth clarifying their exact role. Unlike all the other mixer channels in Logic (which control the flow and qualities of audio within the application), the MIDI channel strips have no direct control of the "internal" properties of your mix, but instead they work as controllers for external MIDI hardware connected to your MIDI interface. However, in the case of the synths and sampler being returned via aux channel strip, these MIDI channel strips could have a direct effect on your audio mix – maybe balancing 16 MIDI channels being returned to one stereo input on your audio interface.

Figure 8.10 A MIDI channel strip.

Master Channel Strip

The master channel strip provides a quick-and-easy access point to the level of signal going to all main outputs and is directly linked to the "volume" control as part of the features on Logic's transport bar. In situations without control room monitor levels, for example, this could be used as a means of adjusting the monitoring level, although it should be noted that any digital bounces made in Logic will be subject to the master channel strip's gain adjustments.

Note that you cannot insert plug-ins across the master fader in stereo mixing (instead use the Stereo Out strip), with the master channel simply being used as a level attenuator. In surround mixing, however, its role – and the role of the outputs – slightly changes, which we'll cover in more detail in Chapter 10.

8.3 Organizing Your Mixer: What You Do and Don't See

Ask any good mix engineers about the fundamental tool that helps them negotiate an effective mix workflow, and they'll often reply "a well-organized mixer." On a traditional console, for example, engineers might decide to repatch the tape returns so that the order and arrangement of the various channels best reflects how they might want to carry out the mix – maybe with the drums in the first 8–10 channels, for example, while bass, guitars, and keyboards following on subsequent channels. So, before you start diving into equalization settings and designing your soundstage, take some time to organize both the arrangement and track list to help navigate your way around the mix successfully. Logic also provides a number of ways to organize and structure the appearance of your mix, for example, separating off sections, hiding unwanted channels, and viewing the complete signal path.

Mixer Views

The first concept to grasp with respect to using the Mixer interface effectively is the link between the Arrange window and the Mixer window. By default, the Mixer is set to its Arrange view out of the three possible view modes at the top of the Mixer window – Single, Arrange, and All. In Arrange mode, the order

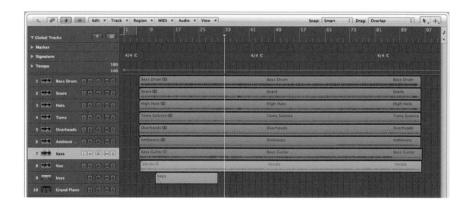

Figures 8.11 and 8.12 In Arrange view mode, the Mixer area directly corresponds to the order of the channels in the track list. The Hide Track option can also be a useful way of removing further channels.

and amount of channels displayed directly corresponds to the Arrange track list. For example, you might have reorganized the vocals to place them on the first track lane – a decision that also impacts on the arrangement of the Mixer. So, one simple way of organizing your Mixer is to organize your track list!

Following this concept, you'll notice that the track Hide feature that we first saw in the editing chapter also carries through into the Mixer. One intriguing difference to the behavior in Arrange window, though, is that the Hide option has an immediate effect on the currently viewed channels, irrespective of whether the H button in the top of the Arrange area is active or not.

An alternative to the Arrange view is the All mode, which displays the entirety of audio channels in your Logic project arranged sequentially by their assignment (Inst 1, Inst 2, and so on), rather than their order in the Arrange area's track list. This is a useful way of seeing the entirety of your mix, including objects (like the metronome and Prelisten channel) that you wouldn't usually have access to. However, without the ability to organize this view, it is easy to become disorientated, especially in bigger mixing sessions.

Knowledgebase 1 ▼

Clipping Faders: Good or Bad?

On the whole, conventional engineering wisdom tells us to avoid situations where a signal "clips" the meters – in other words, whenever we see one of Logic's channels faders or master outputs go into the red. However, in reality, there are some situations in Logic where this is acceptable and others where it demands your immediate attention!

In Logic, a fader registers a clip whenever it is presented with too much signal – this might be because you've brought up the output gain of the compressor too much, or that you're feeding too much signal to the main output channels. Looking first at channel faders and instrument tracks, the clipping we see here isn't particularly problematic, with no immediate chance of you hearing distortion at first. This is all because Logic's Mixer incorporates a degree of safety margin – or headroom – built in, using an enhanced bit-depth resolution to tolerate peaks in the region of 11–16 dB.

Moving onto the main output channel strip, however, things are not quite so flexible. Any final digital medium has a fixed amount of level it can tolerate – in other words, a CD or WAV file cannot exceed 0 dBFS (0 dB on the output fader) without the top of the waveform being clipped or distorted. In this respect, it is important to ensure that the main output fader doesn't ever go into the red when you print off your final mix (although the odd peak

Figure 8.13 Although clips on channel faders aren't too problematic, it is advisable to avoid clipping the main output.

during the mixing process isn't too disastrous). Given the amount of signals potentially feeding this channel, though, this is quite easy to encounter.

So, if your output channel does clip, what should you do to resolve this? The first and easiest solution might be to turn down the faders contributing to the mix, which could be done by placing them all into a group, and then turning the group down accordingly. If it's in the region of a few decibels, you could also choose to turn the master channel strip fader down; although if you're using it in the realms of 2–6 to 2–10 dB, then it suggests that there is a fundamental level mismatch that needs to be addressed. Another option is also to put an instance of the gainer plug-in (on the master output) and also use this to reduce to final mix level ahead of it being bounced.

Viewing by Signal Path

One of the most interesting view modes is the so-called Single mode, which shows the complete signal path of a selected channel. The signal path illustrates the full journey a channel makes through Logic's Mixer, including all the associated bus send effects, any submixing via aux channel faders, and of course, the final output channel strip. Using Single mode is a great way of honing in on the specific signal-processing characteristic of one sound in the mix, especially where it's using multiple bus sends (for reverb, and so on) that could be positioned well along the Mixer in the standard arrangement view.

Dropping Out Sections of Your Mixer

Using the various tabs along the top right-hand corner of the Mixer area, you can switch out various corresponding sections of your Mixer. For example, you might choose to remove MIDI faders given that they don't directly input on the signal processing aspects of the mix. Alternatively, by removing the audio and instrument (Inst) from the view, you can gain quick access to the aux masters and main output without having to scroll across the full length of the Mixer.

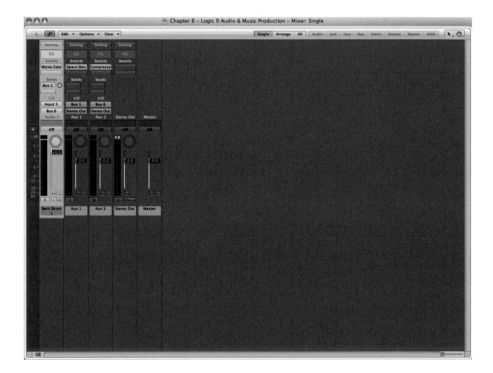

Figure 8.14 In Single mode, you can see the complete signal path for the currently selected track.

8.4 Folders and the Mixer

As we saw in Chapter 4, folders can be useful way of packing your arrangement into groups of related tracks, which can then be viewed and edited independently of the main arrangement. For example, you might decide to pack away your drum tracks into an individual folder, keeping those tracks away from the main view of the track.

Not surprisingly, therefore, the folder feature also has an interesting impact on your view of the Mixer. Any folder tracks are displayed as small strips in the Mixer, without any form of level control, mute switching, and so on. However, by clicking on the small folder, or entering the folder via the Arrange page, you can access this sectioned off part of the mix, using the small hierarchy tab in the top left-hand corner of the Mixer to move backward into the main Mixer.

Managed carefully, folders can be a useful visual tool for keeping on track of the sections of a mix, although, as we'll see later on, there are also other functions for the control of the groups of instrumentation from a sonic perspective. One of the tricks is to decide what you keep on the top level of the Arrange area. For example, if you delete the track lanes from the top "arrangement" level, they'll only be viewable once you've moved into the folder. This can be a great solution for hiding parts of the mix away, although not so good if you still

Figure 8.15 Use the folder function to "pack away" parts of the mix. This can be useful in projects with large amounts of channels or tracks.

want to keep an eye on the entirety of instrumentation. As an alternative, retain the track lanes in the arrangement (even though the regions have been packed away into a folder), only using the folder functionality to "hone in" on specific parts of the mix as and when you want to.

Plug-In Boxout 2 ▼

Helpers ... Handy Little Plug-Ins

Sometimes the small, innocuous little plug-ins can be as vital to creating an effective mix as the big processor-hungry reverbs. Take the Gain plug-in, for example, its main application is simply to raise the gain of an input – surely not that useful then? Say you've recorded in a series of complicated automation moves but simply want to raise the entirety of the track by 2 dB. Usually this would necessitate editing or rerecording the moves. Alternatively, you could use the Gain plug-in (patched somewhere in the insert path) and simply select a 2 dB boost! Some of its other applications can also be great on stereo signals – switching the left- and right-hand sides, for example, or mono'ing a stereo mix (try leaving an instance across the stereo bus) to check its mono compatibility. Also, if you need to phase invert any microphones (a snare bottom, for example), simply activate the Phase Invert option.

The Multimeter plug-in is a great addition to any stereo bus, providing up-to-date information on the track's spectral properties (using a 1/3 Octave Spectrum Analyzer), alongside phase characteristics courtesy of a distinctive Goniometer. Use the spectrum analyzer to get a better grasp

Figure 8.16 Many of Logic's smaller plug-ins can aid the effective workflow of a mix – from the Gain plug-in to the range of metering options.

on the spread and range of frequencies in your mix. If you're using monitors with a limited bass response, the lower spectrum information (100 Hz and below) can be crucial to keeping an eye on any excessive activity. The Goniometer provides some indication of the stereophonic properties of your mix – a strong mono image is indicated by a clear line up the middle of display, while excursions into stereo are indicated by movement to the left and right.

The Correlation Meter – available as a separate plug-in or, on either of the Goniometer or Analyzer's screens – indicates phase problems with the mix. If the phase is good, you should see consistent movement one side of the line (usually in the 11 region). However, if the phase is bad (indicating problematic mono compatibility), the Correlation Meter moves between both −21 and 11 in a continuous fashion.

8.5 Beginning a Mix

With the mechanics of understanding how Logic's Mixer works under your belt, let's start to have a look at the creative and technical processes of putting the mix together. However, before you start instantiating various plug-ins, take time to establish a plan for what you want to achieve. Think about how you intend to distribute instruments across the soundstage – in respect to both left-to-right placement (pan) and front to rear (with the use of reverb and maybe EQ). Establish what you feel should be the "lead" instrumentation – in other words, the three or four instruments that really hold the track together – alongside the other elements that should only provide a "supportive" role. Finally, identify instruments that are possibly fighting against one another, sounds that may blur the definition and energy of the track, or parts that may get lost among other instrumentation. Ultimately, if you're clear about some of these issues from the start, you'll have a much better chance of producing a coherent end result.

Basic Part Leveling

An essential part of the mix revolves around good leveling practice – put simply, balancing parts without the addition of any additional processing. Although the precise technique varies from engineer to engineer, the intention is to get a good "working balance" of sounds, some with more dominant roles than others, but all without overloading the main outputs from Logic. For example, it's easy to start putting a mix together, pushing channels into the 13–16 dB region only to discover that the main output quickly starts to distort. Instead, try to keep an eye on the main output level (usually on the output 1–2 channel, which should sit side-by-side the instrument channel/audio channel in the inspector) and bring in two or three of the key instruments leaving 6–12 dB or so of headroom. Unless an instrument has been recorded particularly "hot," this will probably be in the region of −24 to 0 dB.

Alongside the basic of part leveling, you might also want to consider panning the various channels to create a realistic soundstage. Think about the instrumentation arranged across the stage, creating a natural, well-distributed configuration of the instrumentation, without relying too much on the extremes of left and right (−64 and +63, respectively) and leaving a clearly defined space in the center of the mix for important lead instrumentation like vocals (in other words, only a few channels should be left in the 0 position). Remember that as you apply effects like reverb and delay (which we'll see more of later), you can still pan these signals just like any other channel in the mix.

Applying Equalization and Compression

Two of the most important signal processors in Logic – and the cornerstone of any professional mix – have to be the equalization and compression plug-ins. One of the main applications of compression and equalization is to provide better separation between sounds – fixing different elements into specific parts of the frequency range of a track, for example, or locking a sound better into the mix's dynamic properties. One common technique with EQ is to reduce clashing frequencies between sounds, allowing each sound to sit in its own sonic space. For example, a 200-Hz reduction on a low guitar part would facilitate a little more space for a bass, whereas the intelligibility of the vocal could be improved by a cut at 1–3 kHz on any competing instruments. Compression, on the other hand, helps iron out any dynamic inconsistencies – for example, rather than a vocal dipping in and out of the mix, it can remain consistently loud.

Figure 8.17 Try using equalization to create better separation between instruments. In this example, a 3-dB boost at 3.5 kHz in one guitar has been balanced out by a cut in 3 dB at 3.5 kHz on another.

Plug-In Boxout 3 ▼

Compressor

Along with EQ, compression forms one of the fundamental tools of a mix-down and essentially works as a form of automated amplitude control – the compressor reacts to loud peaks in the input and then attenuates the level of the signal accordingly. The two key parameters in this process are threshold – rated in decibels and governing the point at which compression begins – and ratio, which sets the hardness of compression applied. For example, with a low threshold and ratio (–220 dB and 2:1, respectively), a slight compression would be applied, even on relatively quiet parts of the signal. Alternatively, a higher threshold and ratio (–25 dB and 10:1, respectively) would produce a stronger compression, but this would only be applied on the rare occasion the signal level exceeded –25 dB.

In addition to the key compressor settings (threshold and ratio), Logic's compressor plug-in contains a number of other parameters to further refine the processor. Attack and Release, for example, are important tools for defining how the compressor moves in and out of gain reduction, as the signal exceeds the threshold. With a quick Attack and Release setting, the

Figure 8.18 Logic's compressor is a versatile gain reduction tool, capable of controlling the precise amplitude and dynamic range of a signal in the mix.

(Continued)

compressor is set in a "fast-acting" mode – the response to loud transients (like loud snare hits) is quick and efficient, and the compressor is quick to return to its normal state once the input reduces below the threshold. As great as this sounds, however, these fast movements in and out of gain reduction can have a negative effect on sound being processed, as the compressor creates a distracting "pumping" sound. Setting a slower Attack and Release might allow the odd loud peak to slip through the net, but you'll end up with a more musical, empathetic response from your compressor.

Peak and root mean square (RMS) modes govern how the compressor "listens" to its input. Essentially, the Peak mode responds exactly to the true level of the input, while RMS responds to an averaged level, closer to how the ear perceives loudness. With a Peak detection setting, therefore, you'll find the compressor reacting more than on the corresponding RMS setting, although its response might be considered slightly less musical. The Knee of the compressor is a useful means of creating a more graduated transition on harder ratio settings – the ratio is slightly softer (2:1, for example) ahead of the threshold, only reaching its full strength (6:1) a few decibels after the threshold.

As the overall effect of compression is to reduce the dynamic range of your input (in other words, making the loud bits quieter), you'll probably find your corresponding output quieter than without compression. Increasing the output Gain (at the right of the interface) restores levels lost through compression. You can also use the AutoGain option of Logic to apply this automatically, although in some cases this can produce distortion.

Compression and EQ both can be applied via the insert path of a selected audio channel strips, instrument channel strips, or aux channel strips, although the EQ can also be quickly activated by double-clicking in the small EQ box at the top of the channel strip. Double-clicking the selected plug-in will then open a floating dialog, allowing you to adjust the various parameters contained within the plug-in. Of course, bypassing effects is essential to establishing exactly what you have or haven't achieved. This can be done using the dedicated bypass control in the top left-hand corner of the plug-in's interface or directly from the Mixer itself by option-clicking on the appropriate plug-in slot.

Adding in Further Plug-Ins

Further plug-ins can also be inserted as part of the path of plug-ins you create on the channel strip. Remember, though, that the order in which you insert the plug-ins can have a big effect on the overall treatment produced. Even something as simple as EQ and compression, for example, can have a subtly different

Figure 8.19 The order in which plug-ins are inserted can have a big effect on the end output. Try using the Hand tool to experiment with the order of your plug-ins the channel strip.

output based on which plug-in is placed first in the chain. More extreme effects, say, for example, distortion followed by EQ or EQ followed by distortion, can have completely different result given their relative position. If you need to change this order, you can use the Hand tool on the Mixer (or simply press Cmd + click the plug in) to reorder the inserts in any way you see fit.

One common mistake at this point is to apply too many plug-ins, possibly in a desperate attempt to improve apparent deficiencies in the source recording; "at least if it's covered in distortion and delay, nobody will notice that it's out of tune!" Try to have a clear strategy in your application of effects – not all tracks necessitate the use of plug-ins, and where a more complicated series of plug-ins are used, try to make this a unique and identifiable "special" feature, rather than the norm.

8.6 Adding Send Effects

As tempting as it is to patch everything across a channel's insert points, it's really worth making a clear distinction between the use of true insert-based effects and the use of send effects. The use of an insert effect implies there's no need to balance an unprocessed version and a processed version of the signal. EQ, therefore, is a perfect example of insert effect as there is little or no need to hear both an unequalized version and an equalized version of the sound at the same time. The same could also be said for compression. Reverb, on the other hand, is a clear candidate for being applied as a send effect, with the need to balance the respective levels of wet (in other words, reverberated) signal with the dry, unprocessed version. In theory, the greater the reverb, the more the channel appears to move toward the back of the mix.

Plug-In Boxout 4 ▼

Channel EQ

The Channel EQ is easily one of the most important plug-ins in any mix, controlling the precise spectral qualities of signals passing through it. Although the Channel EQ takes a low drain on available CPU resources, it is both a powerful and a flexible tool, split into eight frequency bands with a cut/boost and frequency parameter for each band. As a good starting point to understanding your input source, try activating the Analyzer feature of the EQ. The Analyzer provides a real-time Fast Fourier Analysis of your signal, indicating the distribution of energy across the sound spectrum – a bass, for example, should produce large "humps" formed by its fundamental at 100 Hz, alongside additional harmonics further up the harmonic spectrum.

The two extreme bands – at the far right- and left-hand side of the interface – govern the controls for the high-pass and low-pass filters within the EQ. Effectively, these completely remove frequencies above or below the given cutoff point – try using the high-pass filter, for example, to tame any excessive low-frequency energy. Moving inward, the next two bands correspond to the shelving EQs, which are somewhat comparable to the treble and bass controls on a conventional hi-fi. Use the shelving EQ for general sweetening activities, creating the familiar "smiling" EQ curve with a boost at around 80 Hz and 12 kHz, respectively.

The remaining bands work as traditional parametric EQs, with a cut/boost control, frequency setting, and a fully variable Q parameter. Q sets the width of cut and boost, and therefore the resultant amount of cut and boost to harmonics is near to the EQ's selected frequency. A wide Q is a useful way of shaping more "general" qualities of the sound – maybe a lack

Figure 8.20 Channel EQ is one of Logic's primary mixing tools: with four fully parametric bands, two shelving controls, and two filter sections.

of bite in the upper mids, for example, or an overall woolliness in the low mids. Use a tighter Q where your frequency issues are more specific – like a boomy resonance on an acoustic guitar, for example, or prominent problem harmonics.

The first step to creating a send from any given channel is to double-click in one of the available send slots on the strip and select from one of the 64 available bus destinations. Raising the level on the small send pot will then bleed an amount of the channel's signal via the bus to the appropriate aux channel strip. The aux channel strip should then have the corresponding plug-in (an instance of Space Designer, for example) patched across its insert path so that signals entering the aux channel strip are affected accordingly. With the mix parameter on the plug-in set to 100% wet, you can now control the level of reverb fed back into the mix, using the aux channel strip's fader or using further plug-ins across the aux channel strip to further process the sound of the reverb.

Both reverb and delay are important tools in defining the spatiality of your mix – in other words, the front-to-back perspective of the soundstage. Having worked carefully on your level and panning, you should have initially formed a good representation of your left-to-right soundstage. However, with the application of delays and reverb, you finally have a chance to set sounds forward and backward in the mix – a dry vocal, for example, sat squarely at the front, or a distant keyboard pad drifting toward the back of the soundstage. One really useful technique is to place two to three contrasting instances of Space Designer to establish a front, mid, and rear acoustic to in your mix. Use some of the shorter reverbs on close sounds, or rhythmic elements (like drums or rhythm guitar), that don't suit long reverb settings. The longer settings, on the other hand, could be reserved for a few unique sounds that really benefit for longer reverb tail and a more distant mix placement.

Figure 8.21 A basic configuration of send effects – the keys channel has been set up to have an auxiliary send delivering a little of its signal through to an auxiliary channel which is configured with a reverb on its 100% wet setting.

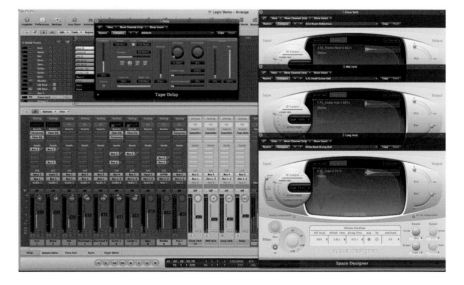

Figure 8.22 Use two or three complementary reverbs to help define the front-to-back perspective in your mix. Additional delays can also help define the "spatial" dimension.

One important concept of the sends is whether they are configured for either pre or post fade operation. In a mix, the most common example (and the one Logic defaults to) is to use the sends in post fade mode – that is, the send happens after any adjustment in level for the particular channel. If, for example, a fader level is brought down in the mix, its corresponding amount of reverb is also attenuated – in effect, preserving the ratio between the two sounds (dry and reverberated). Pre fade sends (as we saw in Chapter 4) are usually associated in the creation of headphone or cue mix. To adjust between pre fade, post fade, and post pan operations, simply click and hold on the send and adjust accordingly.

8.7 Combined Processing Using Aux Channels

Having seen one application for aux channel strips – that of applying send effects – let's have a look at the other ways in which they can be applied during a mixdown. By changing the output option of any audio or instrument channel in the mix, you can also decide to route it to an aux channel strip ahead of the signal reaching any of the main, physical outputs from Logic. Used in this way, you can combine a number of signals – say, the various microphones positioned around a drum kit, for example – to a single fader, with the option to control both its level and the application of additional processing en masse.

Figure 8.23 Bus processing can be achieved by sending a number of channels (via a bus) to an aux channel. On the aux channel, insert the required equalization or compression, for example, so as to process the "group" of sounds.

The use of so-called bus processing across aux channel strips can be a great way of locking together the principal components of a mix, as well as helping groups of sounds gel in an effective way. Many American rock engineers get a great deal of mileage from this technique, combining compression on a channel-by-channel basis, alongside compression across a collection of aux channel strips. The result is a mix that "pumps" in an empathetic way to the source material, providing a real intensity and sense of loudness brought about through the reduction of dynamic range.

However, if you do intend to apply a lot of compression on your mix, it's well worth making sure that you use a variety of different compression techniques as a means of forging the identity between different sounds (rather than the track just sounding squashed), as well as matching the style of compression to the instrumentation you're processing. Try using low ratios (1.5:1 to 2:1) and low thresholds as a means of massaging the sounds into place, whereas harder ratios and higher thresholds (6:1, −5 dB) could be a great way of simply slicing a few loud transients of a drum submix. With any bus compression, you'll also need to pay close attention to the Attack and Release times of the compressor, with slower settings (especially on the release) providing a more musical result.

Plug-In Boxout 5 ▼

Convolution Reverb and Space Designer

Convolution has quickly become the accepted standard in producing professional-grade reverb treatments. The technique works by taking an acoustic sample, known as an impulse response or IR, from a given room or indeed, a hardware reverb unit. The process of convolution then simply takes the short IR file and mathematically "folds it over" the source signal, effectively recreating the effect as if the sound had been recorded in the same space or through the same reverb processor. For that reason, convolution reverb can sound astonishingly realistic, although this can be at the expense of your available DSP resources. Ultimately therefore, it makes sense to run just a few instance of Space Designer, patched across bus faders, than lavish multiple instances on individual channel faders.

As part of the Logic Pro install, Space Designer comes with an impressive Library of prerecorded impulse response files (stored under Library/Application Support/Apple/Impulse Responses). You can load these simply by scrolling through the presets or loading them in via the Load IR option to the right of the IR sample switch. Additionally, you could choose to make use of one of the growing number of third-party IR libraries, like Spirit Canyon's Spectral Relativity, or indeed, your own custom-sampled IR files.

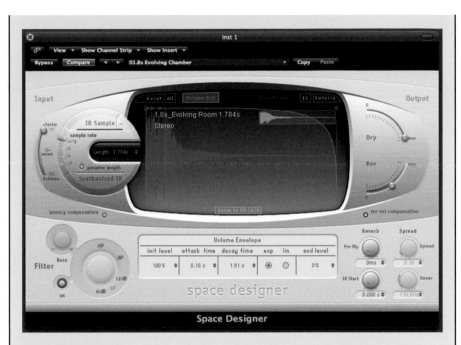

Figure 8.24 Based on unique samples of the original spaces, Space Designer's convolution reverb engine produces some stunningly realistic reverb treatments.

Like a growing number of the professional convolution reverbs, Space Designer offers a surprising amount of control and flexibility with effects you can achieve with it. Once an IR sample is loaded, you can immediately adjust both its volume envelope and its filter characteristics, assuming the filter is activated. Besides being able to produce some subtle modifications (rolling off the high-end, for example, with a touch of low-pass filtering), these tools can be great for abstract ambience treatments – try using the reverse setting and some extreme filter movements with lots of resonance! If you need to shorten the reverb tail, adjust the length parameter. Alternatively, to create a longer, darker reverb, adjust the sample rate into one of its slower settings. This can also be a great way of producing longer reverb times, without maxing-out your DSP resources.

8.8 Using Groups

As alternative to bussing sounds together via aux channel strips, you can also make use of Logic's Group feature – as we've already introduced, with the concept of edit groups in Chapter 5. Taking the concept further, we can also use Groups to lock together collections of channels, say a group of drums, for example, or backing vocals. But given the aux channel strip bussing system, what use is an additional means of ganging faders?

As simple as the bussing system is, there's one major conceptual problem – that of post fade send effects like reverb. Imagine a collection of kit sounds all being sent to an aux channel strip, with a snare also making use of a generic reverb (also shared by the vocals and strings) on send 8. If the level of the kit were then reduced on the aux channel strip fader, you'd hope the reverb level would also be reduced accordingly, but this is not the case. As the send happens before the aux channel strip, any subsequent level changes will have no effect on the reverb – effectively the snare will get "wetter" as the level of the kit falls. Of course, one simple solution to this would be to route the reverb to the same aux channel strip, but this would then have a knock-on effect on the vocal and strings reverb!

So, here we see a perfect example of the application and benefit of grouping channels as apposed to bussing signal via aux channel strips. Grouping retains all the important gain structures of the original mix – the ratios of reverb are retained, even as the level of the group is reduced. Bussing still has its place, though, as there's no way to apply effects like compression and EQ onto the group. So, in theory, a good mix may well involve combination of both grouping and bussing via aux channel strips, grouping to "lock" collections of faders, and bussing via aux channel strips to provide grouped applications of effects!

Looking more closely at the Group Settings dialog, you can see a number of additional features that could potentially improve the speed and efficiency of your mix. Besides obvious grouping controls like volume and mute, you can also link the send controls (this could even be helpful in quickly modifying cue mixes as you're recording), as well as any linked Automation Mode selections.

Figure 8.25
Shared reverbs can make the use of aux channels for grouping purposes problematic; although the fader level of the drums group is reduced, the reverb level stays the same.

Figure 8.26 This example uses both groups and bus processing. The Group allows us to better control the level of the drums in the mix, while the aux channel is solely used for processing rather than level control.

However, one vital keyboard shortcut to use in conjunction with the Groups is the Toggle Group Clutch command. By default, this is configured onto Cmd+G and can be used to temporarily disable all groups – a great way of quickly modifying an "internal" balance within any number of groups, without having to laboriously enable or disable the relevant groups.

Knowledgebase 2 ▽

Parallel Compression

Parallel compression has become a real buzzword in mixing circles, but what is it, and can Logic be realistically used to apply it? Unlike the traditional insert-based approach to applying compression, which works on the assumption that you'll only want to hear the 100% compressed signal, parallel compression offers the unique possibility of hearing both compressed and uncompressed versions. Although parallel compression is arguably less effective at reducing the dynamic range of a signal, it is an excellent way of combining both the hyped-up sound of heavy compression, with the natural dynamics and "air" of an uncompressed recording – yes, to use the cliché, it really is the best of both worlds!

One really simple way of creating parallel compression is to copy the source track onto two adjacent channels. On the one channel, keep the signal

(Continued)

uncompressed and open, while on the other, try setting up really juicy, pumping compression. Don't worry about the integrity of the signal – pick a suitably tough ratio (6:1 or harder!), relatively quick Attack and Release, and let gain reduction pump well into the 6–10 dB range. Some engineers will even EQ this channel, picking out the extreme highs and lows (80 Hz and 12 kHz), as well as applying a little tuck in the mids. With the compression set up, combine the two channels together, mixing in the compressed version to add "balls" and body to the uncompressed track.

Another interesting approach is to use the compressor almost like a send effect – blending any number of sounds through to the same heavy compression setting. Try doing this on a drum mix, sending mainly the close snare and kick drum through to the paralleled compression. Again, pick some suitably pumpy settings, maybe even a touch of extra ambience, and sit the bus in the mix to add the required amount of weight and importance.

Figure 8.27 In this example of parallel compression, a bus send is being used to add compression using the same routing technique as a reverb send. Use a heavy compression setting, and then bleed in the required amount compressed signal to add "body" without compromising transient detail.

8.9 Working with Channel Strip Settings

As you're building up the mix, you may well find that certain channels require similar settings to that of others – maybe you've compressed and equalized one backing vocal, and you want to apply the same setting to corresponding

Figure 8.28 Use the channel strip settings as a quick way of moving one channel's plug-in configuration to another. The presets feature also allows us to do this between songs.

backing vocal track. Clicking on the small arrow to the right of the Insert label allows you to open up the channel strip settings menu. Try using the copy/ paste option as a simple means of duplicating the channel settings on a number of channels. Alternatively, if you have a number of favorite channel strip settings – maybe a particular compression and EQ on a given vocalist, for example – you can also save the presets off, as you would any of the other plug-in presets. These favorites could then be recalled at any point on any given song.

8.10 Automation: The Basics

Despite the considered application of compression, you may still find the demands of your mix changing from verse to chorus and vice versa. Ultimately, although one balance may work at a given point in time of the song's development, the same might not be true as further instrumentation enters and the musical qualities of the track change. In the "golden days" of recording – before computers stepped out of the office and into the studio – it wasn't uncommon for a mix to involve "all-hands-on-deck": the engineer, assistant engineer, and even the members of the band, all frantically pushing faders up and down to shape the mix to the dynamic of the song.

Logic's Automation provides complete control over qualities of the mix throughout the duration of the song – including basics like volume, pan, and mute and more advanced options like plug-in parameter automation. The golden rule to remember with automation is only to apply it toward the ends of a mix – once you've established the principal balance, equalization, compression, and effects usage, only then should you begin to turn to automation. The problem is that any mix changes after automation, although not impossible, can be quite a headache to perform – potentially requiring you to rewrite or delete existing automation moves.

8.11 Track-Based versus Region-Based

One potentially confusing aspect of automation is that Logic contains two contrasting methods for applying automation – track-based automation and region-based automation. In truth, region-based automation – where automation data is contained within the audio or MIDI region itself – is actually a throwback to the earlier versions of Logic and has largely been retained for the purpose of backward compatibility, although, to be fair, it still has its uses. Track-based automation – where automation data is recorded on a separate multiple track lanes to that of the audio region – offers a far more flexible means of creating and editing automation data, as well has having the distinct operational benefit of not being tied to the regions in questions. Editing or moving a guitar solo, for example, won't necessarily disrupt the automation data that accompanies it.

Thankfully, Logic does provide a means of transporting one automation type (region-based or track-based) to another – making the combination of the two

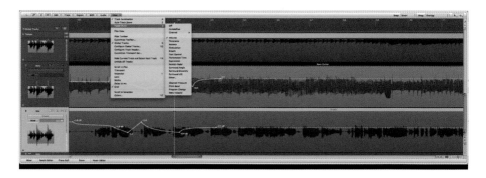

Figure 8.29 Two different approaches to automation – the old-fashioned Hyper Draw (top) and the standard track-based automation (bottom).

approaches quite a powerful feature. For example, you could write the automation using the superior track-based controls and then convert this information into automation data that is stored with the region. For now though, we'll take a look at how track-based automation works and then consider the region-based case later on.

Plug-In Boxout 6 ▼

PlatinumVerb

PlatinumVerb is a valuable alternative to Space Designer, either for Logic Express users (who don't have access to Space Designer) or for situations where you're running slightly low on available DSP resources. In contrast to Space Designer, PlatinumVerb places a surprisingly low drain on the CPU, although correspondingly, you may find its output slightly less authentic than Space Designer. Used carefully however – on short settings with drums, for example, or any instrumentation sitting lower down in the mix – it can produce surprisingly effective results.

Figure 8.30 If you're short of processing resources, or don't have access to Space Designer, PlatinumVerb makes for an excellent alternative source of reverb.

(Continued)

One of the best things about "modeled" reverb, as apposed to convolution reverb, is that it offers precise control over the sound produced in the virtual room. In the case of PlatinumVerb, it divides the reverb up into two key stages – the early reflections, as sound initially bounces back from the walls, and the more diffuse reverb tail, as the reflections merge together to create a definable "trailing-off" to the sound. Try moving the Balance slider over each side to fully understand and audition the effects obtained in each stage.

Using a predominant mix of early reflections can be a great way of thickening up drum sounds by producing a noticeable, distracting reverb tails. Experimenting with the room shape, stereo base, and room size will change the qualities of reflections produced, either by creating a tight centralized reverb or by creating a wider, more expansive set of reflections. When it comes to the reverb tail, try exploring the density, diffusion, and reverb time parameters. On the whole, most conventional reverb effects tend to stay within the realms of 1–2 s, and, not surprisingly, this is where PlatinumVerb seems to sound at its best. Density and diffusion, respectively, govern the spacing and randomness of the taps. Try using both in their lower setting for an effect like Spring Reverb (great on electric guitars), or use a higher setting for a smoother sounding reverb.

8.12 Automation Modes

To engage a channel into automation, you'll need to change automation mode according to the way in which you intend to write automation data into Logic. By default, all faders are set to their Off position, just above the pan pot – this means that any existing automation data are ignored and the faders can be freely moved or repositioned (without fear of Logic snapping them back!) at any point. Engaging any of your faders into an automation Write mode (Write, Touch, or Latch) will allow you to begin writing data based on the current song position. Interestingly, though, the "recording" of automation data is independent of the transport's record switch; in other words, you only need to be in Play mode for automation to be written.

First, let's take a look at the various modes used to record automation data.

Write

Think of this mode as the most dangerous! Any fader engaged into write mode will record data onto the automation lanes, even if there's existing data, so this should really be used with care. Write, however, can be a valid way of deleting and replacing automation data in one pass.

Touch

Touch is the safer way of writing automation data, as it only engages into writing data once the fader has been "touched." Release the fader – even as the track is still playing – and the fader will return the previous recorded position, and carry on reading any existing automation data. Touch, therefore, can be great in creating a few strategic nips and tucks in your mix – briefly either lifting a phrase out for a few seconds or pulling back any part that dominates the mix.

Latch

Latch mode is comparable to touch, in that a fader is only engaged into writing data once it has been touched. However, when the fader is released, it continues to write data – potentially erasing any existing moves – and the level it was last left at. Latch can be a useful mode when you need to raise or lower a level of a part and then leave it at that level for the remainder of the song without having to constantly "hold" the fader in place. Stopping the transport will, of course, stop the writing of automation data.

Read

Read is a "safe" automation mode, where any automation data is read back but no further moves can be written – indeed, if the fader is moved for any reason, it will promptly snap back into place! On the whole, most users tend to leave faders in Touch mode as they're automating, allowing them quickly engage the writing of automation data without having to constantly switch modes. However, to reduce the risk of overwriting certain aspects of the mix, it's best to switch all faders back to Read mode once you have finished automation.

Although the description of different modes has assumed you're working with the channels faders, it's important to remember that once a channel has been placed into Write, Touch, or Latch, any movement in its accompanying parameters – including mute, pan, or any of the plug-in parameters – will also

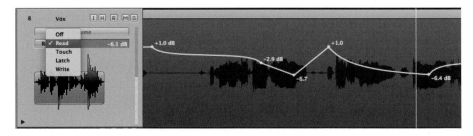

Figure 8.31 Each of the four different automation modes used in Logic has its own impact on how you read and write automation data.

be recorded. This allows any part of the mix to be automated with the same degree of flexibility as the channel faders; for example, the reverb time of a vocal reverb could be swelled going into a "larger" chorus, or the feedback of a tape delay unit could be modulated to create some dub-inspired delay treatments on the end of a line.

Plug-In Boxout 7 ▼

Using Delay

Although most users tend to perceive reverb as the main tool for defining spatial qualities of a mix, it's also surprising to realize just how much spatial interest can be added with something as simple as a delay line. Logic comes with three different delay plug-ins – sample delay, tape delay, and stereo delay – facilitating a range of different delay treatments.

As the name suggests, tape delay is modeled on the type of delay effect produced by classic tape-based delay effects like Roland's Space Echo or the WEM CopyCat delay. As these processors use tape to produce their delays, the effect has a characteristic "dark and dirty" sound to it. By default, the plug-in works with tempo divisions – simply select the required division (semibreve, crotchet, quaver, and so on) and the delay will appear in time with your track, even if the song's tempo changes. Use the Groove slider in the extreme settings (33% and 75%) to change the division to dotted note values. You can also achieve some great slap-back

Figure 8.32 Logic's simple delay plug-ins can be a surprisingly effective tool for mixing – from rhythmic delay effects to subtle forms of "slap-back" ambience.

delays (an effect famously used on early rock'n'roll vocals) by using a semiquaver setting and sliding the Groove parameter down into the region of 33%–50%.

Feedback sends a proportion of the sound back on itself, effectively creating a regenerative delay effect. On the tape delay, feedback settings in the region of 50%–100% will appear to "hold" the delay, with an increasing amount of distortion and grit on each repeat. Stereo delay works on the same principle as tape delay, although it deliberately avoids the tape-based coloration, in preference of a cleaner digital repeat. You can also set different delay time for the left- and right-hand side of the stereo image.

In contrast to the other delay plug-ins, sample delay only deals with incredibly small delays times, measured in samples. As a rough guide, 44 samples equate to 1 ms of delay, so even at its maximum setting (4000 samples), sample delay only gives us about 90 ms to play with! So why use this plug-in? Well, sample delay can be beneficial in applications fixing minute time delays caused when microphones are widely space apart – applying sample delay to correct these anomalies and create a more phase-coherent image. If you're less technically inclined, try using a small amount of sample delay on some drum room microphones to simulate the effect of sound reflecting further away from the kit.

8.13 Viewing and Editing Automation

By default, Logic hides the display of automation data – otherwise an arrangement could soon become cluttered. However, once you've started recording a few moves, you might want to see how the mix is beginning to shape up. Select View > Track Automation to display the current recorded automation – you might also want to toggle this using a keyboard shortcut (the default is A) so that you can quickly switch automation viewing on and off.

With the automation view engaged, you'll notice some important changes to both the tracks and the arrangement itself. Looking first at the tracks, you should now see the automation mode indicated on them (Read, Touch, and so on) alongside the current viewable automation parameter – clicking on this should allow you to scroll through all the parameters available for automation. You'll also see a small bar graph meter indicating the current fader position. Note that this can be freely modified and controlled, just like the "real" fader as part of the Mixer, and can be a great way of adding in a few cunning automation moves.

Figure 8.33
You'll need to enable the automation view mode to see or edit the moves you've recorded.

Where multiple plug-in parameters have been automated, you can also add further track lanes into the equation, simply by clicking the small arrow toward the bottom right-hand corner of the track name in the track list. Alternatively, option-clicking opens as many automation lanes as is required to display all the automation data currently recorded.

Automation data itself is displayed as a series of nodes, which can be manipulated using the usual Pen, Eraser, and Arrow tools. As you'd expect, the Pen tool allows to draw new automation moves, while the Arrow can modify existing node points, or by double-clicking add new nodes into the equation. Indeed, in many situations, it may be quicker and easier to draw in a couple of automation nodes than to laboriously write in a series of moves. Manually placing the nodes can also be a good way of setting in long automation events – a long filter sweep over 30 bars, for example – or when events, like fade-outs, need to happen at a precise time.

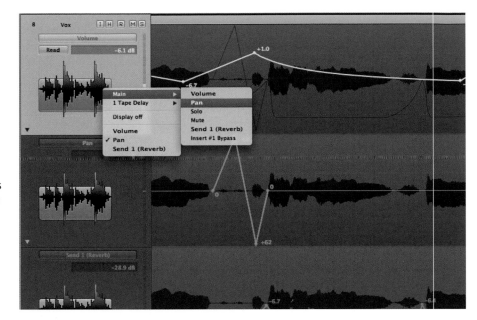

Figure 8.34 Press the small arrow on each track lane to open up a parallel automation lane, making it clear to see the moves on several different parameters on the same track.

Plug-In Boxout 8 ▼

Vocal Processing

Vocals can be one of the trickiest parts of a mix to get right, so it's great to see a range of plug-ins suitable for vocal processing in Logic. Away from the core plug-ins that we've already covered – including EQ, compression, and reverb – you'll also find the DeEsser and Pitch Correction plug-ins helpful in crafting that polished vocal performance your mix requires.

Logic's DeEsser is a handy way of taming any excessive sibilance (like the letters s and t) that might have been accentuated by a poor choice of microphone. The DeEsser works by analyzing the input and then applying a selective amount of gain reduction whenever problematic sibilance is heard. The Detector part of the interface is what should be used to spectrally locate the sibilance in the vocal. Set the Monitor mode to Det (detection) to tune into the specific frequency your signer's sibilance is occurring (usually somewhere in the region of 6 kHz). Now, flick to the Sens (sensitivity) monitor setting and adjust the sensitivity so that it only flicks on when the sibilance occurs. Moving the monitor to its OFF position, adjust the Suppressor parameter to the same parameter to that of the Detector and increase the strength to get the required amount of sibilance reduction.

(Continued)

The Pitch Correction plug-in is loosely based on the infamous Antares AutoTune plug-in and provides an on-the-fly means of correcting intonation problems in a vocal – or indeed, any other monophonic performance. At extremes, using the response setting on its 122 ms setting, the plug-in can be forced to produce the clichéd, quantized vocal effect so carelessly abused in the late 1990s! On softer sensitivity settings however, it can be a useful way of taming any problematic pitch drifts. To achieve the best results with the plug-in, you'll need to specify the key and scale the song is in. A number of presets for this are available, but you can also manually switch the notes in and out by clicking on the accompanying keyboard.

Figure 8.35 Logic's DeEsser and Pitch Correction plug-ins can aid a number of problems in relation to your vocals.

When two nodes are placed manually, you can also make use of the unique Automation Curve tool as a means of adding a degree or "curvature" to the line. Simply click and hold on the line between the nodes, using the Automation Curve tool, and drag above or below the line, or from side-to-side, to create one of four different adjustable curve shapes.

Where ranges of moves need to be moved or duplicated, you can also drag-enclose a number of nodes simply by holding Shift as you rubber band a group of nodes. With the nodes selected, they can then be moved en masse, or with the option key held down, duplicated to a new position. One other potential lifesaver, in case you've written in the automation moves but feel the need for them to be a couple of decibels higher or lower, is Logic's scaling feature. To scale a track's automation moves, use the Command key and drag up or down on the small bar graph meter as part of the track list. Look carefully at the selected values, and you should be able to spot the corresponding changes you've made.

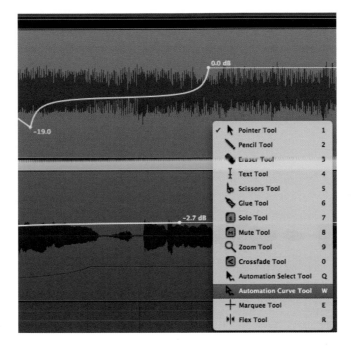

Figure 8.36 Use the Automation Curve tool to create neat and natural curves between automation modes.

Knowledgebase 3 ▼

Saving CPU: Freezing and More ...

At some point in the mixing process, it's highly likely that you'll start to hit dreaded inevitability of a CPU overload. Dealing effectively with this problem could make the difference between a half-baked mix and a distinctive mix, using appropriate strategies to save CPU resources without compromising on sound quality. First, it helps if you're clear as to which plug-ins are the real CPU monsters in your session. For example, synthesizers like the ES2 and Sculpture can take up a big quota of resources, while the EXS24 (assuming it not making lots of use of its filter) is surprisingly processor efficient. In respect to plug-ins, Space Designer can be really CPU-hungry, especially if you're using longer IR files in excess of 2 s, while the Channel EQ is almost negligible on today's Intel-powered machines.

Arguably the first technique, therefore, is to look for any optimizations you can apply to the session that might improve overall CPU efficiency. For example, try turning off unused oscillators in the ES2, or removing excessive Space Designer use on inserts, in preference for a few instances on aux sends.

(Continued)

The next step is to use the track freeze option. Freezing effectively creates an audio bounce of the track in question, complete with all its inherent plug-in settings, and then deactivates the plug-ins accordingly. If you then try to edit any part of the track – either opening an instrument or plug-in or repositioning a region – Logic will remind you of its "frozen" status. If you do need to carry out the edits, simply unfreeze the track, make the modifications, and then refreeze.

Freezing itself is carried out with the small freeze icon as part of the track header. If you can't see the icon, select View > Configure Track Header and then add the freeze button into the set of options. Note that the freeze occurs across the full duration of the project, so it's worth moving back the project end marker to the real finish point of your track.

As an alternative, you can also carry out an audio bounce yourself, simply rendering complex virtual instruments as audio regions, saving off the instrument settings, and packing away the MIDI data (just in case you need to go back). This can be an effective solution earlier in the project where you intend to do lots of structural rearrangement with the regions at a later point.

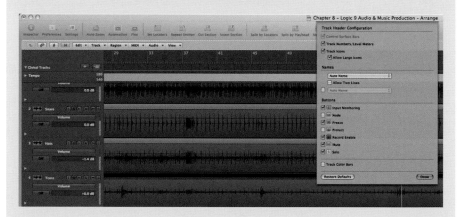

Figure 8.37 Use the Freeze function to render an off-line version of the track and so release valuable CPU resources.

8.14 The Automation Menu Options

Besides the various graphic tools for editing automation data, there's also an accompanying automation menu (Track > Track Automation) containing various options to delete moves, as well as some intriguing options to move data back

Figure 8.38 The Track Automation menu includes several powerful features, including the ability to move automation between its native track-based form and the region-based solution of Hyper Draw.

and forth between track-based automation and region-based automation (as previously discussed). Although it is, of course, easy enough to delete selective parts of automation using the eraser tool, the menu options make quick-and-easy to create "blank-slate" on individual automation lines, tracks, or indeed the whole song itself. Certainly, in situations where you might inherit a previous song file for a new project, this can be a great way of clearing out problematic automation moves.

Moving track-based automation to region-based automation is useful in situations where you'd like the automation permanently attached to a part, without having to continually specify for the automation data to be moved every time you realign or duplicate the part. Possibly the best "real-world" application of this would have to be the use of automation to control filter movements – maybe you've written a couple of distinctive filter movements that are as much part of the musicality in the region as the notes contained within the MIDI sequence. By switching the automation over to region-based (Track > Track Automation > Move Visible Track Automation Data To Region), you effectively lock the movements into the part, allowing yourself to forget about whether the data are moved or not. Of course, at any point, this same data can be brought back to track-based automation using Track > Track Automation > Move Visible Region Data to Track Automation.

Knowledgebase 4 ▼

Compression Circuit Types

The qualities of the different compressors are something that tends to excite many professional engineers, mainly as the different approaches to circuit design can achieve some radical differences in the types of compression achieved. Ultimately, this means that certain compressors tend to suit themselves to specific applications, bringing plenty of "character," as well as gain control, to the input they're processing. In the example of Logic's compressor plug-in, therefore, we're provided with Platinum, ClassA_R, ClassA_U, VCA, FET, and Opto.

Although the exact models from which these circuit types are derived aren't supplied, it is worth noting some of the key differences between these different options and how best to apply them. Opto represents the oldest compressor design, based on early models that used optical cells as part of their gain control circuitry. This unique design resulted in a degree of latency – both with respect to the Attack and Release on the compressor – that tends to deliver a more "musical" compression, as apposed to a harder, more aggressive gain control. The Opto circuit type, therefore, works well with bass sounds (that don't suit fast Attack and Release times), vocals, or anything that you want to retain a degree of musicality and lightness with.

Figure 8.39 The different circuit types approximate the unique sonic behavior of many classic types of compressor like the Urei 1176 or LA-2A.

The FET design, on the other hand, used Field Effect Transistors to create a more heavy-handed compression, particularly good at catching transients. The FET circuit model therefore, and to our ears the ClassA_U, can produce some really effective results on drums, especially when used across overheads. In this application, don't be afraid to use low threshold setting with "pumping" Attack and Release times for an aggressive, almost low-fi compression sound.

Although this provides a theoretical background to select the circuit type, the best approach is to use your ears. Try configuring some basic compression settings and then flick between the different compressor models to hear the marked differences in how they sound.

Walkthrough ▼

Adding Compression

Step 1:

Insert a compressor across the instrument you want to process and start by establishing some basic settings. Working from the default positions, try finding a ratio and threshold setting that works for the instrument you're

Figure 8.40

(Continued)

trying to process. For example, for a gentle compression, use a medium threshold with soft ratio (1.5:1 through to 2:1) yielding about 2–3 dB of gain reduction (another term for compression). For a harder compression effect, consider bringing up the ratio (4:1 or more) and lowering the threshold to achieve 6 dB or more of gain reduction.

Step 2:

With these basic settings established, you can now start to refine the compression a little. Try adapting the Attack and Release settings to best suit the style of compression you want and the sound you're trying to squash. Slower settings (Attack 40 ms, Release 400 ms) tend to create a more natural progression in and out of gain reduction, although you might find the occasional loud transient slipping through the net. Faster Attack and Release settings (Attack 0–10 ms, Release 100 ms) produce a "pumping" effect, which tends to work well where you want the compression to sound more noticeable.

Figure 8.41

Step 3:

As compression leads to an overall loss in level, you need to raise the output to restore the overall signal to its original peak level (although, of course, the signal will be more compressed). Try bypassing the compressor, noting the meter readings, and then using the Gain parameter (with the compressor active again) to restore the original level. Listen carefully to the compressed sound in the mix – is there enough compression to sit the instrument correctly? Is the compression too obvious? Further fine-tuning (maybe increasing the ratio or softening the Attack and Release) optimizes the compression for the instrument's position in the mix.

Figure 8.42

Logic Tips ▼

Plug-In Delay Compensation

Any plug-in added into a channels signal path creates a small amount of delay or latency, through the extra processing required in producing the effect. Thankfully, however, Logic includes a feature to compensate for any delays incurred through plug-in processing, called plug-in delay compensation (PDC), available under the general tab of the audio preferences (Preferences > Audio). For users using standard audio unit plug-ins (either Logic's own or from other developers), stick to the "audio tracks and instruments setting." However, if you are running a processing-accelerator system like Universal Audio's UAD-1 or TC Electronics' PowerCore system, you may notice delays building up when you start to use buses to apply UAD-1

(Continued)

or PowerCore plug-ins. In these situations, change the preference to its All setting – Logic should now play the entirety of the session in time.

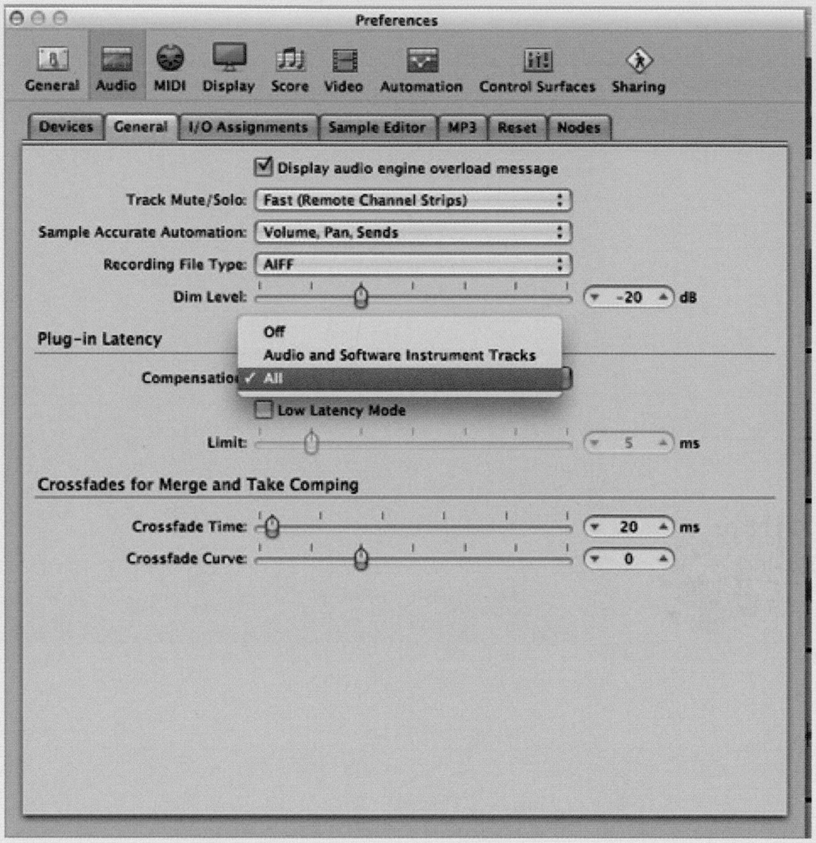

Figure 8.43 Changing the PDC settings is important for users of processing-accelerator systems such as Universal Audio's UAD or TC Electronics' PowerCore.

Plug-In Boxout 9 ▽

Pedalboard

Pedalboard is an ideal companion plug-in to Amp Designer and is designed to replicate the array of footpedals that guitarists might use to shape the sound of their guitar. As such, you'll probably want to place Pedalboard ahead of an instance of Amp Designer, although of course, you're free to use it anywhere along your signal path should you see fit. Indeed, the combination of both Pedalboard and Amp Designer shouldn't just be reserved for guitars, as the two plug-ins can work wonders on a range of sounds – from synth basses, to lo-fi breakbeats, and vocals!

Pedalboard is easy to use – simply drag the required pedals from the Pedal Browser on the right-hand side of the interface over to the pedal area found on the left-hand side of the plug-in. The signal path works from left to right, with each subsequent pedal adding the previous pedal's output. This means the order of the pedals can have a big effect on the eventual output, although it's generally best to place 'tone' pedals, like compressors, equalizers, and distortion pedals, earlier on in the signal path; with effects pedals, including flange, chorus, delay, and spring reverb later on in chain of footpedals.

As well as using a left-to-right 'serial' routing, Pedalboard also allows you to create parallel routing configurations where you can split the signal between two separate busses and then sum the results at the end of the chain of footpedals. For example, this makes it possible to layer two different types of distortion, rather than each distortion unit adding to the results of the previous pedal, or to send a split a sound to be processed by reverb and delay as separate entities.

You can assign a pedal to Bus B using the Routing area above Pedalboard – simply click on the pedal you want to move, and it switches to the upper Bus. Note a Mixer object now appears at the end of the signal path, allowing you to blend and pan the two discrete signal paths (A and B). Note that you can also manually insert a Splitter pedal to achieve similar result, although the added benefit here is that the Splitter pedal can also divide the sound by frequency, allowing each bus to process a different part of the frequency spectrum.

Figure 8.44 Pedalboard replicates the array of footpedals guitarists might use ahead of their amp to shape the basic sound of the guitar.

In This Chapter

9.1 Introduction   309
9.2 Different Approaches
    to Mastering   309
9.3 Bounce to Disk   311
9.4 Audio Mastering in Logic   314
9.5 Editing Fades   314
9.6 Exporting and Burning   317
9.7 Mastering in WaveBurner   320
9.8 Processing and Editing   322
9.9 Dithering, Bouncing,
    and Burning   324

Knowledgebases

Pre-Mastering   312

Dithering   313
Disc Description Protocol   332
POW-r   333

Plug-In Focus

Linear-Phase EQ   315
Multipressor and Multiband
    Compression   319
Limiting   323
Other Mastering Tools   324

Walkthroughs

Editing and Assembling a CD in
    WaveBurner   328
Audio Mastering and CD Burning   330

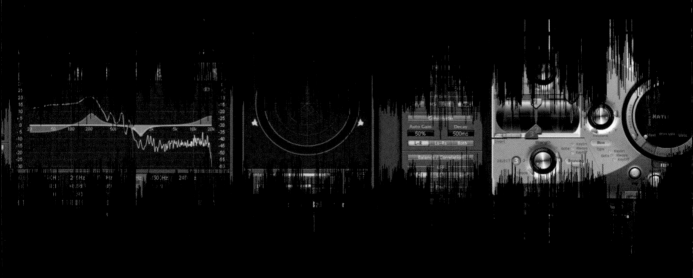

Mastering in Logic

9.1 Introduction

If you've ever tried burning a few tracks onto a CD, you'll be well aware of the challenges of producing a CD that sounds comparable to a commercial release. Even with a complete mastery of the production process, and some great-sounding final mixes, your CD could still sound weak and, comparatively, amateurish. Although OSX and iTunes offer integral CD creation and burning, the fact is that professional musicians and bands will invest a significant amount of money and experience in turning their finished mixes into a final product. So does this mean that users of Logic can't enjoy the same degree of finesse and polish? Well, with a little know-how, and the audio tools of Logic, you too can produce a release-quality CD master.

The term mastering describes the process of compiling and editing several (possibly contrasting) recordings, applying some form of audio "sweetening," and assembling these to produce a final Red Book "production master" CD. Traditionally, mastering has necessitated the use of a separate facility (other than the music studio), specially equipped with mastering equipment including multiband compressors and high-end mastering EQ, as well as dedicated workstations like Sonic Solutions, Pyramix, or Sequoia. Nowadays, however, the world of media is a lot more demanding, so it's not uncommon for musicians, bands, and composers to master by themselves. The bar, it appears, has been raised – but Logic is certainly up for the challenge!

9.2 Different Approaches to Mastering

Mastering itself can be divided into several objectives. First, of course, is the sweetening we most commonly associate with a commercial CD – in other words, the use of compression, EQ, and a host of other processes across the finished two-track master. Second, tracks will also need to be edited – setting

DOI: 10.1016/B978-0-240-52193-0.00009-5

correct start and end points for example, or placing any fade-ins or fade-outs as required. Finally, the finished files or regions need to be ordered for the CD, with appropriate markers to define the tracks and index points that appear on the CD. With these three objectives, you can master in the Logic universe.

Technique 1: iTunes

The first technique, and the one most Logic Express users are used to, is the idea of applying mix sweetening, edits, fade-outs, and dithering all options inside the main Logic application. The finished files are then rendered (using Bounce to Disk) or exported as 16-bit 44.1 kHz files ready for compilation. The compilation process, however, requires the use of other software – either iTunes (Apple's integral audio-CD burning tool) or dedicated Red Book software like Jam.

Technique 2: Burn from Logic

For the quickest and the most integral solution to the mastering problem, you can burn CDs directly from Logic. Again, sweetening and edits can be applied directly to the main application. The burning of CDs, however, is carried out from the Bounce menu – rather than rendering an audio file, you'll burn the finished mix onto a CD. This method is, arguably, the quickest way of creating a

Figure 9.1 Audio Masters, exported from Logic, can be assembled into a finished CD using iTunes or any other suitable Red Book standard software.

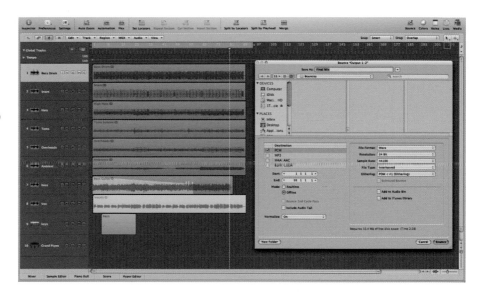

Figure 9.2 For a quick CD, use Logic's integral CD Burn feature, as part of the Bounce to Disk option. Although this method is an easy way of burning single tracks (a rough mix, for example), it can be too restrictive for professional mastering.

CD, especially with just one or two tracks needing to be burned, but it doesn't offer the most flexible solution in the long run.

Technique 3: WaveBurner

The best technique – although only available to users of Logic Studio – is to use Apple's dedicated Red Book application called WaveBurner. Originally developed by Emagic, WaveBurner has previously been sold as an individual program, but is now included as a standard in Logic Studio. In essence, WaveBurner shares many features with Logic (including plug-ins like Multipressor or Denoiser) alongside tools specifically dedicated to the process of assembling a Red Book standard CD. The real advantage, however, is the way in which you can experiment with the order and sound of your CD – with automatic track crossfades, individual plug-ins for each region (or track), and the important Red Book options like International Standard Recording Codes (ISRC) and CD-Text.

9.3 Bounce to Disk

Whether you are mastering in Logic itself, or using another application (like WaveBurner), the first step will be to render your Logic mix as an audio file. The temptation to apply audio sweetening or fades in the main Logic project can be persuasive, but this should be avoided at all costs. Ideally, if you have at least one copy of your song unmastered at 24 bits, then it would be more suitable to take to a professional mastering engineer (when the lucrative record deal arrives!) than a home-mastered, 16-bit file. Separate high-resolution files also

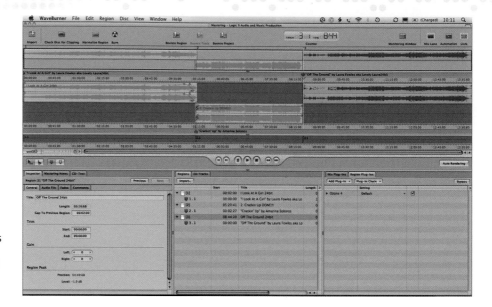

Figure 9.3
WaveBurner offers the best method for mastering Red Book compliant CDs.

afford you the opportunity to approach mastering your work away from the "headspace" of a mix, so that tracks sound correct in terms of the whole CD rather than on an individual basis.

The Bounce dialog window can be accessed through the Bounce button, on the main output channel strip. Before clicking on this, you'll need to define the length of the bounce, designated by the current cycle length – without this, Logic will simply default to bouncing from the beginning of the first region to the end of the last. By defining the length manually, you can keep the best account of a few crucial factors – namely, the "hangover" at the end of the track generated by reverb tails and the allowance of a small amount of silence at the start of the audio file. At this stage, it's probably best to render a slightly larger file, rather than too small, as trying to add information later (say, for example, when a reverb tail gets cut off) can be tricky, if not impossible.

Knowledgebase 1 ▼

Pre-Mastering

Officially, the term "mastering" refers to the process of cutting a master disc from which the records would be pressed in duplication. Pre-Mastering originally described the process that engineers used to prepare the audio signal for the vinyl medium. Nowadays, the term mastering tends to be used in place of Pre-Mastering and describes the process of preparing the audio for the intended medium, ordering the music and also signal processing.

In the Bounce window, you'll need to specify the file type, resolution, and dithering options. Ideally, a PCM, AIFF, 24 bit, 44.1 kHz, interleaved file is considered the best "raw data" for mastering. More important, you should check that the dithering has been set to none, as dithering is best applied at the very last stage of mastering when the word length is reduced to a 16-bit master. The bounce itself can be carried out in Realtime (maybe you've got some live synths or effects coming into the audio mixer) or Offline – a quicker way of rendering to file by temporarily devoting all your computer's resources to the bounce process.

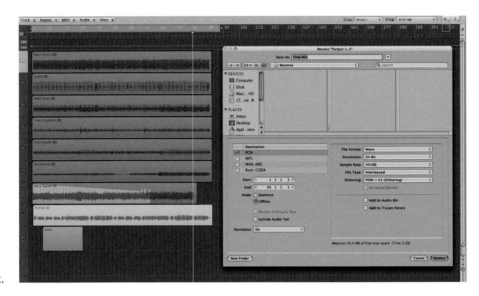

Figure 9.4
When performing a bounce in your Logic arrangement, check that the duration of the bounce is slightly longer than your track and the resolution is 24 bit.

Knowledgebase 2 ▼

Dithering

Dithering is an essential part of the analog-to-digital conversion process. However, the process of dithering a signal also refers to the process of altering the word length to suit your specification. For example, if you bounce your Logic song as a 24-bit file but wish to place this on a Red Book CD, you will need to dither this down to 16 bits. The 24-bit recording will allow for a much more accurate 16-bit file because of the increased resolution, unlike the "dither" noise added in 16-bit A/D conversion. Logic employs a licensed algorithm called Psychoacoustically Optimized Wordlength Reduction (POW-r) from the POW-r Consortium LLC (http://www.mil-media.com/docs/articles/powr.shtml). The POW-r algorithm is considered one of the best dithering algorithms and is widely supported by mastering engineers around the world.

9.4 Audio Mastering in Logic

With the raw data of your tracks assembled, you can begin to look at mix sweetening. Whether you're using WaveBurner or Logic, the principal plug-ins and objectives will be the same, although the exact details of their application will vary. With the various songs pulled into a master Logic session, you could place each region on a different track and experiment with various plug-ins to reach the desired sound. Ideally, the CD should present a rounded and uniform tone throughout, with a consistent "loudness" across the tracks – you could even try importing some commercial tracks as a reference to see just how far you can take things. Exactly how you achieve this will vary from track to track.

The principal tools used in mastering are well represented in Logic, with a number of dedicated mastering plug-ins, including phase-linear EQ, multiband compression, and limiting. Obviously, you can use any plug-in where appropriate, but these particular tools will be the most useful in achieving a professional sound. Plug-ins can be inserted either on the individual track's insert points (for song-specific processing) or across the main master-channel strip for general application (maybe when the album needs to be limited as a whole, for example). As with any audio processing, the order of the plug-ins is vital for the end result, although the widely accepted order for mastering is EQ, followed by compression, and finally limiting. Metering (applied across the main output using the Channel EQ's Spectrum Analyzer or the Multimeter plug-in) will help keep an overview of things – note, in particular, how the commercial tracks might meter differently to your own mixes.

9.5 Editing Fades

Besides preparing the sound of the tracks, it's also important to look at some other important mastering details, like the exact start and end points, and any desired fade-ins or fade-outs. Having already mastered audio editing in

Figure 9.5
Mastering in Logic using a combination of channel inserts and the main stereo output's insert points.

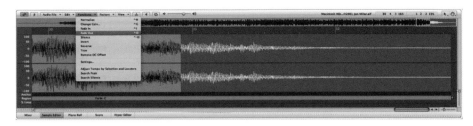

Figure 9.6 Use Logic's Sample Editor to perform basic "top and tailing" tasks, like silencing unwanted noise or creating small fades (as illustrated).

Chapter 5, you should have no problems switching over to working on whole songs rather than individual regions. When you're editing the start point, however, make sure you leave a small amount of silence and don't cut right to the beat – this allows for older CD players to demute and stop the track from "jumping-out" at the start of playback. Also, pay close attention to unwanted noise at either end of the track – the easiest way to address this is to either use the Sample Editor's Functions menu to "Silence" unwanted noise, or use a quick Fade In and Fade Out as appropriate.

Creating proper track fade-outs (i.e., fading out over the last chorus) is a little more taxing. The problem lies in the fact that adding a fade using Logic's

Plug-In Focus 1 ▼

Linear-Phase EQ

Having looked at Logic's standard channel EQ in the mixing chapter, let's take a look at the mastering-orientated Linear-Phase EQ. Being much more CPU intensive than the Channel EQ, the Linear-Phase EQ produces a technically superior sound by removing the phase shifts (caused as frequencies are cut and boosted) that occur in a conventional EQ. Even with its CPU drain, the Linear-Phase EQ is a welcome and sonically accurate tool for enhancing and shaping the timbre of a track in mastering. Operationally speaking, the Linear-Phase EQ includes the same controls as found on the Channel EQ, and the same phenomenally useful FFT analyzer.

Applying EQ in mastering requires an approach different to that of mixing. On the whole, your approach needs to be as subtle as possible – remember everything you do will be much more noticeable as you're only

(Continued)

Figure 9.7 The Linear-Phase EQ removes any of the usual phase shifts that occur with conventional parametric equalization.

working with two stereo tracks! Try to keep boosts, or preferably cuts, to a maximum of +/-3 dB with a wide bandwidth (low Q parameter) – if you're going beyond this, you may well have a problem with the original mix. Overall, your track should exhibit a smooth response, with a gentle roll-off of high frequency energy. Remember though that there are other tools that can also have a timbral effect on the mix – sometimes more successful than EQ – like multiband compression (especially on bass) or harmonic excitement (for some top-end sparkle).

conventional method – arrangement area fade tool – doesn't account for the use of mastering processors. As the fades are preeffects, any corresponding change in the level of the track will result in a change in compression – so as the track fades out, for example, it slowly loosens its compression. To counteract this, try performing another bounce, this time with the processing in place. The new "compressed" region can then be reinserted into Logic (the old effect being disabled), and then, the fade can be applied.

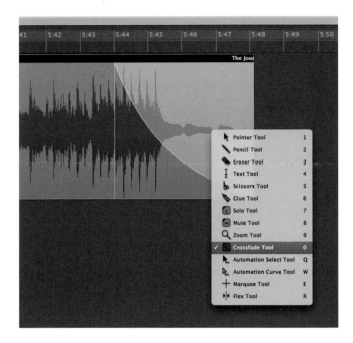

Figure 9.8 For longer fade-outs, you'll need to render the file with effects, and then create the fade using either Logic's region fade tool (as illustrated) or the Sample Editor's fade.

9.6 Exporting and Burning

Once the tracks have been processed and edited correctly, you can begin to create the final exported 16-bit masters ready to be burned, or indeed, you could burn the files directly from Logic. The Bounce dialog window again comes into action – either to create the final rendered files or provide access to the Burn feature. Given the regions of the exact finished song length, you can use the Region menu to select "Set Locators by Regions," and in this way, the cycle length (and therefore the bounce) will be exactly the same length as the region. The Bounce dialog window should specify the creation of 16-bit files, which will necessitate the (final) application of Dither to smooth out any quantizing noise brought about by moving away from 24 bits. Once exported, these files can be dragged straight into iTunes (or other suitable software) ready to be burned.

With burning now supported directly from the Bounce dialog window, you can also burn a CD directly from Logic itself. Again, where the session contains audio files greater than 16-bit resolution, the addition of dithering is essential, which can be selected from the appropriate pop-down menu in the Bounce window. Other pertinent options include setting the write speed (keep this low to avoid any write errors that degrade the audio quality on playback) and writing the disc to a multisession CD. When the "Write as

Figure 9.9 Using the "Set Locators by Regions" feature you can create a new bounce with exactly the same length as that of your edited regions.

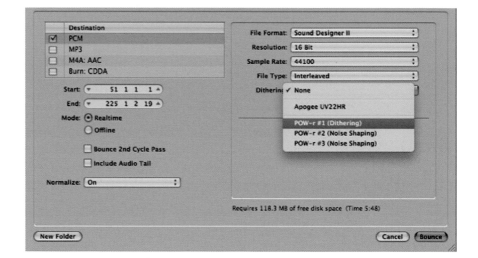

Figure 9.10 The last bounce will need to be dithered down to 16-bit resolution, ready to be burned in another application.

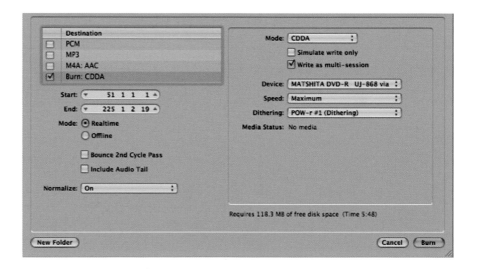

Figure 9.11 Quick and easy CD creation: Logic's Bounce and Burn feature.

multi-session" option is selected, you'll be able to add further mixes to the CD at a later point, otherwise Logic will simply burn a single file onto the CD. Although adding sessions at a later point will save the use of CD-Rs, you might experience problems with CD players recognizing songs burned in subsequent sessions.

In many ways, the "Bounce and Burn" feature is best viewed as a quick fix for creating audio CDs – especially if you intend to write more than one track to the CD at any given time. In this scenario, it might be, perfectly, legitimate to attempt some rough mastering in the actual project's file, rather than a separate session – simply insert a Multipressor and some EQ across the main stereo outs and away you go!

Plug-In Focus 2 ▼

Multipressor and Multiband Compression

Originally developed as part of WaveBurner, Multipressor is Logic's answer to the big multiband compressors (like the TC M6000 or TC Finalizer, for example) so frequently used in mastering. Unlike conventional single-band compression, a Multiband compressor splits an incoming audio signal into separate frequency bands before applying compression. Separated in this way, a complete mix is far easier to control and gel to form a finished master. In most cases, the greatest amount of compression will be applied to the bottom end of the mix – keeping the bass solid and tight – with the high-end requiring a light and minimal touch.

Multipressor features up to four bands of compression and downward expansion, although, to make things easier, you might just want to use three bands (low, mid, and high, respectively). One of the most important things is to set the right crossover frequencies for the different bands, as this can have a big effect on the finished compression achieved. To set the crossover point, try moving the vertical borders on the top of the interface's window – you could also use the Spectrum Analyzer (part of the Multimeter plug-in) to visually analyze the track's constituent components, as well as the band solo feature.

Now, work your way through each band, adjusting the relative compression ratios, thresholds, and attack and release settings. On the whole, most compression ratios tend to fall below 3:1, with the most productive results achieved in the area between 1.5:1 and 2:1. Adjust your threshold to achieve the correct amount of gain reduction and dynamic control.

(Continued)

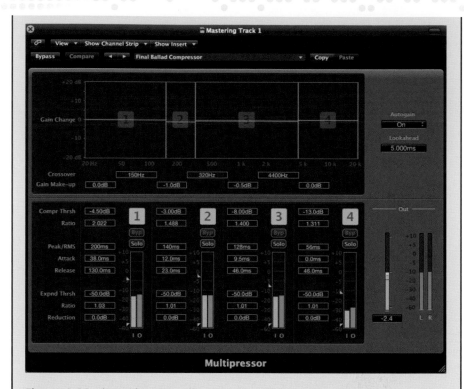

Figure 9.12 The Multipressor allows you to compress your master using a series of different frequency bands.

Again, mastering is often distinguished by a lightness of touch, with typical amounts of gain reduction rarely exceeding 2–4 dB. Attack and release times change the responsiveness of the compressor – avoid setting the release too fast (the compressor might start to pump) or squashing the transients too much with a fast attack. Finally, rebalance the bands using the Gain Make-up control. If you've compressed a band particularly hard, you might need to bring its level up to restore its position back into the mix.

9.7 Mastering in WaveBurner

For a fully professional result that meets the requirements of Red Book standards, you'll need to use WaveBurner. Preparing for this will simply involve rendering the files (ideally at 24-bit, without dither) ready to be imported into a WaveBurner session – after this, the rest of the mastering process (compression, editing, and so on) can be carried out in WaveBurner's domain. The advantages offered by WaveBurner stem from the fact that it is designed, from ground-up, as a tool for mastering – Logic, however, is a dedicated production tool. Track crossfades, for example, can be difficult to create in Logic – this is not the case in WaveBurner. Additionally, with support for CD-Text, UPC/EAN codes, and index marks, WaveBurner is one of the most complete Red Book compliant applications available for the Mac.

After creating a New session (File > New), raw master files can be inserted in WaveBurner through the Region > Add Audio File menu, or by dragging the audio file from Finder. WaveBurner's screen is divided into four main areas, the most important being the Wave View and Overview areas (toward the top of the screen) that provide a graphic representation of the CD and the various regions that comprise it. The Region list, toward the bottom right-hand corner of the screen, lists the series of regions (or audio files) used in your session – a

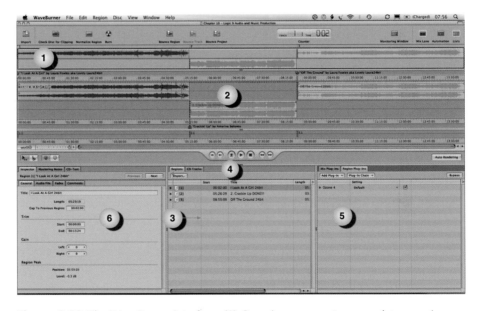

Figure 9.13 The WaveBurner interface. (1) Overview – presents a complete overview of all the regions currently used in your WaveBurner session. (2) Wave View area – gives a more detailed presentation of the regions, also where you can perform the various editing tasks (fades, trimming, and so on) in WaveBurner. (3) Regions List – a list of audio files used in your session. (4) Track List tab – changes the pane to show the order and spacing of regions in the playlist and forms the structure of your CD. (5) Plug-ins list – displays the currently configured plug-ins used for the selected region, or the "global" plug-in inserted across the mix outputs. (6) The Inspector – allows you to tailor the audio and glean information about your files. In addition, there are tabs to add Mastering Notes and CD-Text.

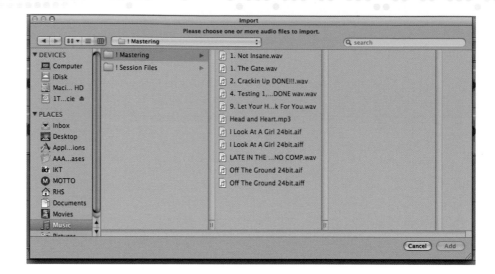

Figure 9.14
Importing new audio files that are ready for mastering.

song, for example, could be constructed from several regions stuck together. The Track list, adjacent to the Region list, lists the Song marker used to break up the CD – more important, these can be completely arbitrarily placed and aren't necessarily tied to the regions.

9.8 Processing and Editing

Audio processing is carried out in the Region and Mix plug-ins list. These work in much the same way as the channel and the master inserts did in the explanation of mastering in Logic. So, for example, the Region plug-ins can be used to process audio files imported into WaveBurner on a song-by-song basis, whereas the Mix plug-ins are applied to the entire program output – in other words, all the tracks. Although the Mix plug-ins can be useful for processing en masse, they are best used for metering and audio analysis. WaveBurner includes the full set of Logic plug-ins, as well as support for third-party audio units.

With WaveBurner's specific focus on mastering, the editing tools work more quickly and more effectively than Logic's Sample Editor. Changing start and end points, for example, results in WaveBurner, intelligently, shuffling the other tracks in line with your edits. Fades can be quickly addressed by adjusting the nodes at either end of the region, with options to change the fade curve. More important, unlike trying to master in Logic Pro, these fades are carried out after the Region plug-ins, ensuring a consistent tone irrespective of the fades. Dragging regions over one another creates a crossfade – either to iron out edit points in the two track masters or to create smooth transitions between different tracks.

the use of phase in achieving extra width – instead, it uses a frequency distribution system to pan alternating frequencies to the left- and right-hand speaker, respectively. By using this frequency distribution method, Stereo Spread avoids the phase problems often associated when a width enhancer is put back into mono. In truth, its effectiveness is somewhat limited, and it only works best with material possessing little or no stereo information.

Figure 9.16 Although they have little everyday use, Logic and WaveBurner's other mastering plug-ins can be useful solutions to a range of mastering problems.

Denoiser uses some clever trickery with FFT filtering to produce a cleaner audio signal in situations where the master possesses a large amount of unwanted noise. Like a conventional noise gate, Denoiser needs its Threshold set carefully to make the best use of the effect – try locating a quiet section of the master, with the noise present, and set the Threshold just above this. Reduction defines the amount of Denoising taking place, although if pushed too hard the Denoising can sound almost as distracting as the noise itself! The Noise Type fader seems a little strange at first, but it does make sense – in the center position, the fader is sensitive to noise across the entire audio spectrum, toward the top its bias is toward darker noise, and to the bottom it is more sensitive to high-frequency noise.

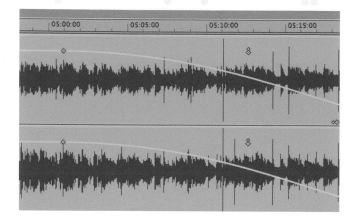

Figure 9.17
Creating fades
in WaveBurner
completely with
a variety of fade
curves. Drag the
small nodes to
define the length
and shape of
the fade.

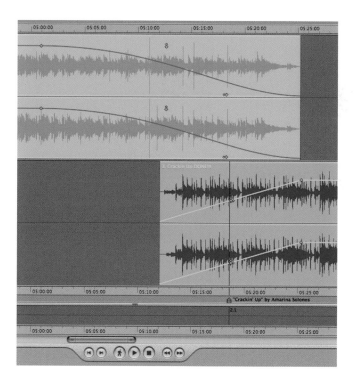

Figure 9.18
Crossfades are
made by dragging
regions into each
other's timeline.

What isn't immediately obvious is that the Bounce preferences also dictate how
WaveBurner burns audio to a CD – particularly with respect to the application of
Dither. Therefore, as long as the Bounce preference is set to dither, WaveBurner
will automatically dither your files as part of the burn – without specifying the
need to burn. In the rare case that you are importing files already processed

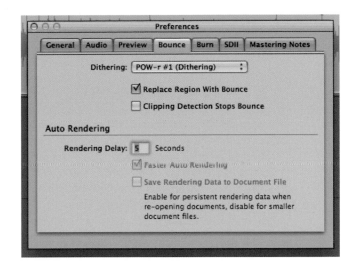

Figure 9.19
The Bounce preferences contain WaveBurner's dithering options – applied either to a bounce or to the final burn.

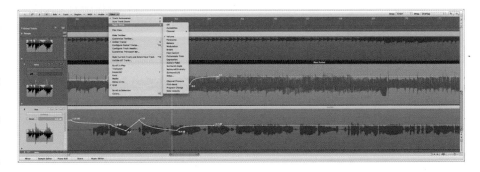

Figure 9.20 Use the Bounce Project icon to render WaveBurner's output as an audio file, or to create a contiguous image file of the final CD.

and dithered, you'll need to remember to return to the preferences and turn the dithering off.

By pressing the "Create Disc icon," you can initiate the burn – although this is only the beginning of the process of learning mastering. For something so apparently simple (how difficult can two tracks and a multiband compressor be?), the real art of mastering professional CDs will take a lifetime to master. Ultimately, the art of mastering is a surprising blend of a gut instinct for what sounds right (in other words, the mastering engineers "ears"), alongside a complete understanding of the technical criteria of manufacturing an audio product. At least with Logic and WaveBurner you have some powerful tools in hand, and the opportunity to present your music in the best possible way.

Walkthrough 1 ▼

Editing and Assembling a CD in WaveBurner
Step 1:
Start a new project in WaveBurner (File > New) and import your required 24-bit session masters (File > Import Audio File). Try resizing the three areas of WaveBurner's interface (Overview, Wave area, and Regions lists) to best suit your needs – here, we've selected a small overview and Regions list, alongside a large, clear wave area for editing. With the songs imported, you'll probably need to perform some basic edits before going any further. To do this, click on the region you want to edit (from the overview) and click and drag either the start or the end point to resize the region. Leave some duration of silence at the start to allow for CD demuting.

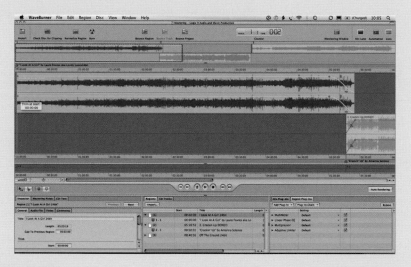

Figure 9.21

Step 2:
Add fades onto either the beginning or the end of the track by dragging the small node at either end of the region. You can change the curve of the fade, using the other two nodes that appear – sometimes the linear fade can be a little too obvious to the ears, whereas curves can be more gentle or extreme depending on the material you're playing with. Another important thing to look at are the gaps and pacing between the tracks, as illustrated by the purple-shaded "pause time" area. Try slipping the next region backward or forward to match the required pace of transition – this is an important way of setting the feel of the album.

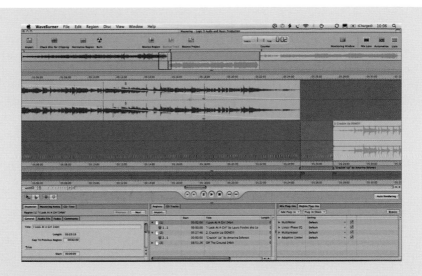

Figure 9.22

Step 3:
Where a region is dragged back beyond the shaded pause area, a crossfade can be created between the two tracks. As with the fade-in and fade-out, the curve of this can be adjusted to get the smoothest, most musical transition between the two tracks. WaveBurner will automatically place the Song

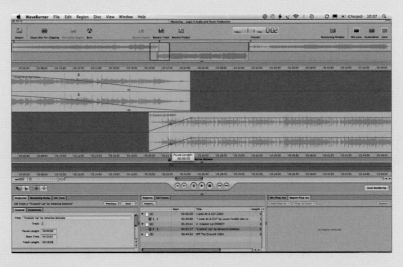

Figure 9.23

(Continued)

marker halfway between this transition – if you want it placed elsewhere, drag the small purple marker on the top of the wave area back to the appropriate point. In addition to song markers, you can place index points – simply change the marker tool (bottom right) from purple to orange.

Walkthrough 2 ▼

Audio Mastering and CD Burning

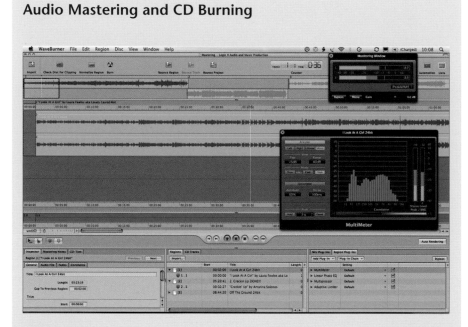

Figure 9.24

Step 1:

With the structure of the CD defined, you can turn your attention to the sound in an attempt to get a "loud" consistent tone throughout. To perform this, some informative metering is essential so that you can hear and see detailed qualities of the track you're working with. In the Mix plug-ins list, insert an instance of WaveBurner's Multimeter and the Level Meter – this should give you a good overview of the timbre and the overall level of your CD. Try importing a commercially mastered track (similar to the CD you are working with) to get some reference to how things should look and sound.

Step 2:

Now turn your attention to the sound of the individual regions. Select a region from the Region list, click on the Region Plug-Ins tab, and insert the appropriate processing to achieve the desired sound. In most cases, a combination of the Linear-Phase EQ, Multipressor, and Limiter (in that order) should pull the mix up to the point where it sounds "loud and proud." Check the results against the other tracks so that you achieve a consistent tone. Most important, avoid excessive clipping or distortion, which you can identify through careful listening, or (from a technical perspective) using the Disc > Check Disc for Clipping menu option.

Figure 9.25

Step 3:

If everything's sounding sweet, you're ready to do the final preparation for the burn. If your source files are 24-bit, you'll need to check the dithering status of WaveBurner. To check this, go to WaveBurner's preferences and select the Bounce tab. Now pick a dithering option from the pull-down menu – POW-r # 1, 2, or 3. Dithering is applied either when you Bounce files in WaveBurner, or as it creates the final burn. To set the burn in action, click on the Create Disc icon and insert a blank CD-R into your CD writer. To ensure a minimal error rate, use the lower burning speeds.

(Continued)

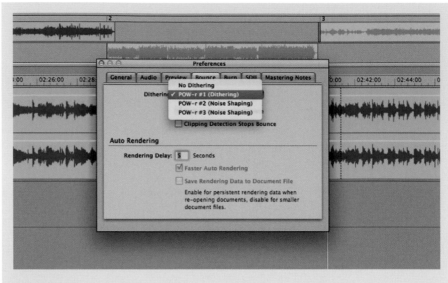

Figure 9.26

Knowledgebase 3 ▼

Disc Description Protocol

Until Logic Studio 9, WaveBurner was unable to write Disc Description Protocol (DDP) images. This protocol, developed by Doug Cardon and Associates, is a way of describing the whole audio CD as simply data. Until now, only top flight mastering applications allowed for this. The introduction of this feature within Logic's suite of applications finally means that the whole production can be handled within your Mac.

DDP images can be written to a CDR-R, or a DVD-r, dependent on file size, but can be placed in a folder for sending over the Internet. DDP offers a degree of quality control not always available from a simple burned audio CD. As the DDP is essentially data, it follows high-standard error correction, and as such it is unlikely to work if it is in the slightest bit corrupted. However, if it works, it will be perfect, and this ensures that the quality of your master reaches the pressing plant in perfect condition.

To write a DDP image from WaveBurner, simply go to File > Save DDP Image... This will create a few files, so you may wish to create a folder for the image. Do ensure that you take the time to complete the Mastering Notes section in full before writing your image.

Knowledgebase 4 ▼

POW-r

POW-r comes with three options within Logic. One is labeled "Dithering" which employs a noise-shaping curve to reduce noise. The remaining settings offer two types of noise shaping. The Logic Manual describes the first noise-shaping algorithm, which can "extend the dynamic range by 5–10 dB." The second noise-shaping algorithm is intended for work with speech as it can "extend the dynamic range by 20 dB within the 2–4 kHz range – the range the human ear is most sensitive to" (Apple Logic Manual). These noise-shaping curves have been developed to be sympathetic to the human ear and draw upon considerable research beginning with Fletcher and Munson's Equal Loudness Contours (More information on Equal Loudness Contours can be found in Rumsey, F. & McCormick, T. (2002) *Sound and Recording, An Introduction.* Focal Press).

In This Chapter

10.1 Introduction   335

10.2 Managing Movies 335

10.3 Global Tracks   339

10.4 Synchronizing Logic   348

10.5 Score Editing and Music Preparation   352

10.6 Surround Sound in Logic   358

10.7 Delivery Formats   365

Walkthrough

Creating AC3 Files Using Apple Compressor   369

Knowledgebase

Surround Sound Plug-Ins   363

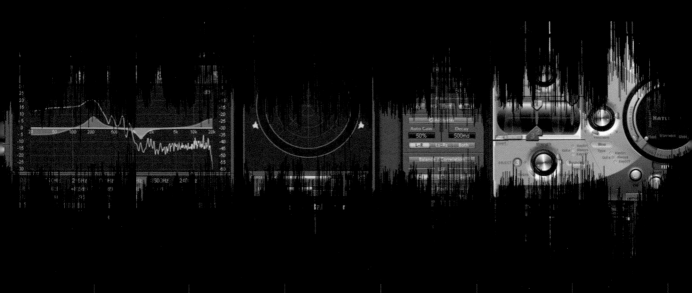

Logic and Multimedia Production

10.1 Introduction

Writing music to picture for film and television often required a whole host of equipment from some form of video player synchronized to your computer sequencer and Digital Audio Workstation. However, Logic takes this task in its stride, and today, whole soundtracks can be written for picture from the one application. This has been made possible by the thorough integration of a visual player and high-level synchronization options from the ground up. This flexibility has meant that Logic has become one of the most popular products for film and television music production.

Additionally in the past few years, we have seen a sudden rise in other outlets for the music producer, such as gaming and Internet multimedia. Logic, with its integration of codecs to export audio as MP3, AAC, and many other formats, has positioned itself as a key production tool for many industries other than recording and production.

In this chapter, we take a movie file and begin integrating it into the Logic project and explore how to write effectively for film and television. We'll look at working with surround sound, markers, scoring, and the delivery formats expected from the industry.

10.2 Managing Movies

Logic has been at the forefront of music and audio composition for picture for sometime, and as such, there are many ways you can work with visuals. You can, of course, synchronize Logic to an external Visual Editor such as Avid or Final Cut, if you so wish, but you're more likely to obtain a video file as an .avi or .mpg (Quicktime) ideally with embedded timecode so that you can work wholly within Logic itself.

DOI: 10.1016/978-0-240-52193-0.00010-1

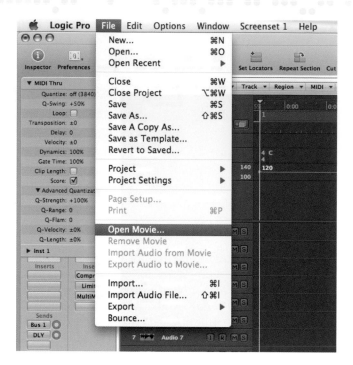

Figure 10.1 To get started on your composition, you'll need to select Open Movie from the File menu.

Before starting work on the music and audio for your visuals, you will need to import the movie into Logic. To begin to integrate movies within your song, go to Open Movie (File > Open Movie…). A second method involves selecting your movie using the Open Movie button in the Video lane within the Global Tracks, which is covered later.

Once you have located your movie in the browser, it will appear as a floating window, which sits neatly on top of Logic. This can be moved around to fit the screen arrangement you have currently. This floating movie window can be resized to suit, and this is achieved by Ctrl + clicking the video window. This produces a new small selection box that reveals itself to offer you the choice of an alternative size for the window or the Video Project Settings.

The Video Project Settings allow you to control a number of things starting with where the movie presents itself. Within the Video Output menu, it is possible to select the preferred output. This will naturally remain as "Window," which will open the movie in the standard floating window. For using a second monitor attached to the DVI output of your Mac, select Digital Cinema Desktop and then select the size of Video Format for that screen. This monitor cannot be used for any other function while Logic is in this mode. Other options include the professional DV output associated with the picture industry called DVCPRO HD and FireWire for sending to an external video interface.

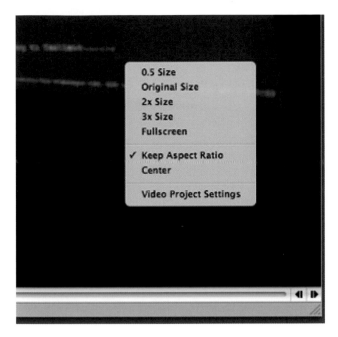

Figure 10.2
Ctrl + clicking the movie window opens up many options such as the size of the movie window itself.

Figure 10.3
The small movie viewer in the Inspector can be a real space saver when working on a smaller screen.

The audio from the original movie can be piped through to the main output, although you may wish to control this by importing the audio into the Logic project, which we'll cover later. Other flexibilities are available here, such as the movie start. For example, it is often unlikely that you'll want the video to start running from 00:00:00:00, but may indeed need some preroll for any introductory credits that might not have come to you yet.

If you're working on a small monitor or a laptop, Logic has a neat way of monitoring what is going on in your movie. At the top of the Inspector, there will now be a movie header. Simply click on the triangle on the left-hand side to open up a small movie viewer.

The Follow Tempo option fixes the speed of the video to the tempo of the Logic song. This link is set by something called the basis tempo. If the project tempo alters, then the speed of the video can relate to the difference between this and the basis tempo just like an Apple Loop would speed up to follow a new project tempo. To the bottom right of this Project Settings pane lies another button called Video Preferences.

Launching the Video Preferences pane reveals some more settings that can be adjusted for our work with video. The uppermost sliders, Video to Project and External Video to Project, allow for the video and timeline to be adjusted against each other, thus altering the start point of the movie if perhaps some lip synchronization is slightly out from a provided dialog track.

When working with video, Logic displays stills of your movie in Global Tracks (covered later). To allow Logic to do this, the computer needs to hold the movie in something called a Cache. A Cache is a form of buffer that holds some of the required data in memory for quick access. These settings allow you to specify both the Cache resolution of the video information and the maximum Cache size allowed. By altering these, you can manage how much video detail you can store in the Cache and thus the response speed when scrubbing.

Extracting the Movie Audio Content

Quite often, you might be given a visual that has integrated audio, perhaps a dialog track for the scene you are composing to. It will be prudent to extract this dialog track for you to be able to manage its level and mute throughout

Figure 10.4
Importing guide audio from the movie can make embellishment a little easier.

tempo, in conjunction with the accompanying visuals, could instantly move your audience to another emotional space.

Changing the tempo is very simple and can be achieved in the Global Tracks. Either select the pencil tool (Escape > 2) and click the tempo change at the desired place or choose the Tempo tab from the Lists pane, which can be accessed by the icon on the right-hand side of the Arrange window.

Using the Pencil tool can offer you a quick method to try ideas out in the project. Clicking the Global Track will reveal a new tempo, which can be glided up or down to instigate a tempo change. You'll notice that the way in which these tempo changes are represented look very similar to that of the automation lanes and as such can be manipulated in similar ways.

For example, you might choose to introduce the piece at a slower tempo and increase it gently. To achieve this is easy, simply set the two different tempos at the ideal point. The jump will be instantaneous and will sound odd. There is a small point in blue at which the tempo changes from the original tempo to the new tempo. This first point can be moved backward in time to introduce a more subtle curve to the tempo change, thus making it gradual.

The menus within the Tempo tab offer some more options to tempo management. For example, you can view the subtle steps that Logic has placed in the project to increase the tempo. To do this, you need to click on the Additional Info button to the left of the Edit menu in the Tempo tab. Prior to selecting this feature, only the start and end point of the tempo change can be seen.

Figure 10.7
Tempo changes can be altered quickly and flexibly in the lists pane on the Arrange window or by editing the Tempo Lane within the Global Tracks.

With the steps expanded, it is possible to quickly edit any step to suit. As you edit the tempo of one of these discrete steps, the Global Track will reveal the steps graphically on screen. However, to create a more staggered, creative, or editable tempo change, simply choose Options > Tempo > Tempo Operations, which can offer some more options to the on-screen Global Track.

A new pane appears that allows you to edit more precise and in-depth tempo changes and alterations. There are many different editable features here, such as the types of curves and their resolution. Within the Operation menu, there are a number of really useful features that could save time, such as stretching the curve out or simply scaling it a little.

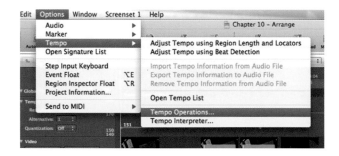

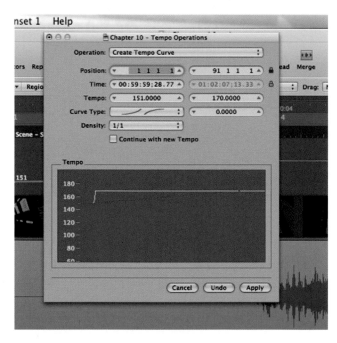

Figures 10.8 and 10.9 The Tempo Operations dialog gives some additional control over making tempo alterations.

Figure 10.10 Adding a key change in the Global Tracks is easy by double-clicking in the lane where a dialog box allows you to specify the change.

The signature lane in the Global Tracks is split into two; the top is for the time signature and the other for the key. As with the Tempo lane, there are again two main methods of changing the signatures. The first is through the lists on the right-hand side Lists window, which contain a tab dedicated to signatures. Within this tab, there are two buttons, one for Time Signatures (create signature) and the other to Create Key changes.

The second entry method is to click again on the lane itself with the Pencil tool (or simply double-click with the pointer). Whether it is for a signature or a key change, a slightly different dialog box greets you. The first of these is the time signature that not only allows you to simply select the signature you want but also gives you the opportunity to enable beat mapping, which we covered in Chapter 5.

Adding a key change is really useful for quick ideas and changes and will work with Apple Loops and any audio you have recorded in this Logic song. The key signature here will enable the Loops to alter in real time to your changes. Simply select the point at which the key change is supposed to occur and double-click (or click one using the Pencil tool), and a dialog box will emerge requesting the desired key and also whether you wish to disable double flats and sharps.

Alternatives

It is often difficult to be sure that the choices you make are definitely going to be right for the outcome of the visuals. Often, it is nice to make changes and try out ideas without losing the original version. Logic has an interesting solution that allows different experiments.

Each Global Track has something called Alternatives, which offer different perspectives on the same track. Similar in nature to alternative playlists when quick swipe comping which we covered in Chapter 5, the Alternatives menu in the Global Tracks allow you to make some different decisions without losing your original root idea. Simply click on the Alternatives menu and choose an alternative number. It is useful to remember that by pressing Alt as you click on your new choice, the data are copied from one alternative to the other.

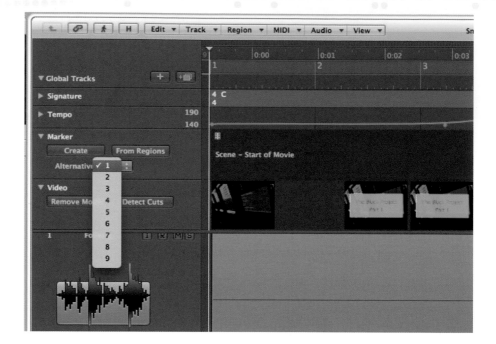

Figure 10.11
Alternatives are useful to try out different permutations of ideas in Global Tracks.

Big Displays

Working to picture is always dependent on timecode, which governs the synchronization between different cameras and audio recorders for editing purposes. Timecode typically comes in the form of SMPTE, which splits the time into divisions of Hours:Minutes:Seconds:Frames, and in some cases, subframes.

The transport bar always shows the SMPTE time of the playhead in the top left with the accompanying bars and beats below. When spotting sounds or effects to film, it will probably be necessary to see a larger timecode display. Ctrl-clicking the transport bar brings up a menu from which you can choose from the Big Bar Display or Big SMPTE Display. These are displays that remain within the darker section of the transport bar. There are also two options to show these as separate floating windows by selecting the Open Giant Bar Display and Open Giant SMPTE Display, respectively (shown on next page).

Spotting Audio

Working to picture will from time to time mean that you will be given a spotting sheet from your clients listing the sound effects. This will tell you where all the sounds need to occur to a particular timecode point and may also include an out time point. These points are given in SMPTE timecode, and in Logic, it is easy to

Figure 10.12
The Customize
Transport
Bar dialog
box provides
considerable
flexibility in the
information and
control you have
over your project.

Figure 10.13 The playhead's SMPTE position can be specified by double-clicking the timecode readout in the transport bar.

edit to these points. It's worth sometimes working to a snap value of "Frames" when working to picture to ensure that each frame is synchronized with the audio.

Two main ways of spotting audio are offered. First is by using the Move to Playhead Position feature. Simply set the playhead to the point you wish for the audio to be placed. This can be best achieved by double-clicking on the SMPTE display in the transport bar and typing is straight in. Pointing toward the audio region, click with control pressed down to reveal the contextual menu and within this will be the Move to Playhead Position command.

The second and more likely method is to edit using the Event tab from the Lists window on the right-hand side of the Arrange area. Presuming you have called up all your audio regions onto the Arrange area, they should be listed here. To hone in on the parts you wish to spot, ensure that Link is on both the Events List and the Arrange window, and simply click on the region you wish to view. The Events List Tab is naturally set to show bars and beats and will need to be changed to view SMPTE through the View Menu (View > Event Position and Length in SMPTE Units).

To spot the audio, simply double-click on the position column to type in the desired timecode. The audio should in theory be spotted, although it might be likely that you will need to delve in further to more accurately ensure synchronization with the picture. Nudging the regions by frames can be a really quick way of making things fit. As we mentioned before, it is often wise to work to frames for the time being, although there will be times where you wish to work in finer resolutions. To change the Nudge Value, right-click on any region and select Set Nudge Value. Once set, simply select the region to be moved and select either the Nudge Left or the Nudge Right from the region's contextual menu or press Alt + Cursor Left or Alt + Cursor Right.

A really fast way to spot audio is to position the playhead at the desired point and select the track you wish the audio to be placed. Next, open the separate Audio Bin Window using Cmd + 9 or Windows > Audio Bin. Select the audio file in question pressing Cmd as you click, and the audio file will be placed at the appropriate place. If you then select another audio file in this way, it will be butted up to the last audio file. There are a couple of useful key commands that can be brought in here called Pickup Clock (move Event to Playhead Position) and Pickup Clock and Select Next Event. These can be set in the key commands menu (Alt + K).

10.4 Synchronizing Logic

Working to picture or on larger projects might involve the need to link Logic to another player. For movie work in the past, it had been often the case that the video would be on another machine, perhaps tape-based, and Logic would need to synchronize to it. Similarly, there may be times when Logic needs to synchronize to another multitrack system, whether that be a legacy open reel 2-inch 24-track tape machine such as a Studer A800 and Otari MTR90 or simply another ProTools rig on another computer. Either way, Logic has a suite of features to cope with your synchronization needs.

Logic syncs in a number of different ways that are industry standard. The main system used for video is SMPTE, and it is an audible code that can also be translated into a digital equivalent used in MIDI called MIDI timecode (MTC). MTC is a popular system used between MIDI devices, and in the latter example above, it is likely that this would be the protocol used to connect a ProTools rig to your Mac with Logic on.

In either case, you will require a device that understands the timecode protocol. All MIDI interfaces should understand MTC, but only some devices are able to translate audible SMPTE timecode to MTC such as MOTU's MIDI Time Piece.

Getting Locked

To tell Logic that it will need to slave to another timecode signal requires you to select the Customize Transport Bar preferences by Ctrl + clicking the transport bar. Select the Sync check box that will generate a button on the transport bar, which looks like a clock with a large arrow facing into it.

In an ideal world, the synchronization should be as simple as selecting this button, and wait for an incoming sync signal to get the project moving. However, invariably you will need to edit the synchronization settings to suit the project you're working on. To do this, either Ctrl + click the Sync button on the transport bar and select synchronization settings or vising File > Project Settings > Synchronization.

Within this pane, lie a number of key features for working with external synchronization sources. The frame rate of the synchronization is very important.

Figure 10.14 The Sync button seen here in blue can be selected from the Customize Transport Bar options. With this selected, Logic works in slave mode to incoming synchronization signal.

Figure 10.15 Right-clicking the Sync button on the transport bar gives you very quick access to the pertinent settings such as synchronization source and settings.

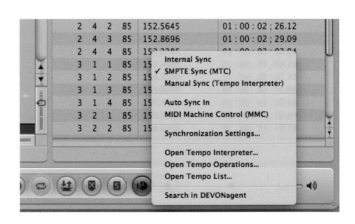

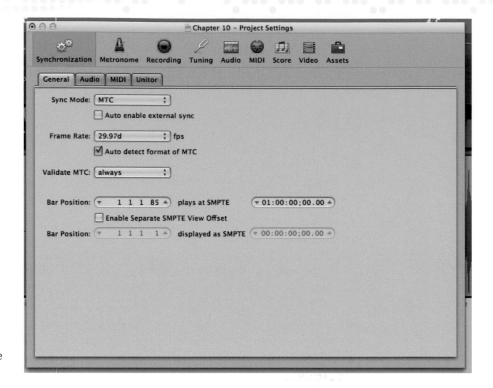

Figure 10.16
Use the synchronization settings to select frame rates and the usual offsets needed from time to time.

Historically, frame rates differed depending on where in the world you worked; for example, 30 frames per second is used for television in the United States of America, whereas we worked with 25 frames per second in the United Kingdom. This was a simple division on the frequency of the electricity supplies in these regions at 60 and 50 Hz, respectively.

Strictly speaking, we're no longer governed in quite the same way, and Logic can work at whatever frame rate is required. Within this menu are considerably more frame rates than we've hinted to above and it will all depend on the project you're working on as to what frame rate you'll need. Safe to say, Logic can handle them all!

Within the synchronization settings pane are some opportunities to offset the timecode to the arrangement. These offsets are very important when working with film. For example, you may wish to use a fresh Logic project for a Scene 1 hour into a film. It would be rather silly to start the Logic project at 1 hour also because things like Freeze tracks would take a considerable amount of time to process!

Setting the offset is easy. Simply select the SMPTE time at which your song should play. For example, Logic will open with Bar 1 playing at 01:00:00:00:00,

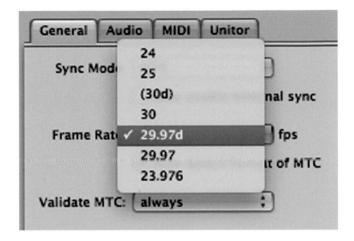

Figure 10.17 The frame rate will need to be matched to your synchronization source to keep up!

which means 1 hour. SMPTE reads as Hours:Minutes:Seconds:Frames:Sub-Frames. However, there will come a time when you will need to alter this to suit. In the example above where we'd like to start the new Logic project at 1 hour into the movie, we'd need to set an offset of an extra hour. As such, we'd want the SMPTE reading here to read 02:00:00:00:00. Logic will then offset the incoming timecode to allow Logic to start playing at 2 hours.

Below the main offset selection is an option to allow the bar positions to show the absolute timecode reference, which can be useful when working to external sources. It is important, for example, that the timeline you work to reads the same as the incoming timecode. Due to the offset in the example above, the Logic project will show a time of 00:00:00:00:00 despite being an hour into the film. This can be changed here, so the project starts Bar 1 at 1 hour, but also shows an accurate portrayal of the timecode it is receiving.

Being the Master

Logic also has the ability to be the master and to allow other devices to slave to its timecode. In this instance, it is necessary to return to the synchronization settings as outlined above and click on the MIDI tab. Here, we can see a checkbox titled Transmit MIDI Clock, which refers to the tempo settings of the project. This clock pulse can be outputted using the MIDI protocol to connect other devices such as effects units whose delay settings can respond in time with the project. Alternatively, in rare occasions, this can be used in part for synchronization.

Below this is the Transmit MTC checkbox and output menu. Here, it is possible to select a device (or "All") to send the MTC to. This will enable any device connected using MTC will follow Logic. For example, if the ProTools rig was to sync

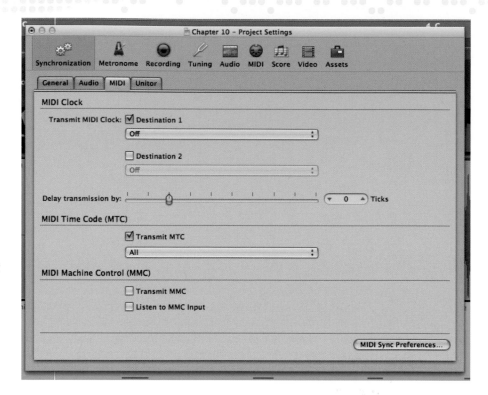

Figure 10.18
Transmitting MTC can be engaged by selecting an output from the Transmit MTC menu, or leave it to send out of all available MIDI outputs.

as a slave to Logic, then simply connect a MIDI cable between the output of Logic's interface to the other computer with ProTools' MIDI In.

MIDI Machine Control is a remote control protocol that enables connected equipment to be controlled by another device. For example, Play on the master could be engaged by controls on a remote slaved device. This can still be a hugely beneficial option to employ across large studios.

10.5 Score Editing and Music Preparation

Although the Score editor can be used for MIDI editing activities in Logic, its most useful features lie in the "preparation" of music – in other words, taking the raw MIDI information that you've performed in the project and transcribing this as finished, musical notation. To be fair, though, Logic doesn't compete with the publishing-standard output of dedicated industry-standard scoring programs like Finale or Sibelius, but its features are more than adequate either to produce parts for a small-scale overdub session, for example, or in the case of a full orchestral session, a sensible "intermediate" format to present to a proper orchestrator.

The process that we're going to explore here, therefore, is the rudiments of taking an existing MIDI composition and turning that both into a full score for the conductor and parts for the musicians. Rather than being a complete exploration of the scoring features and the art of orchestration (which is a book in its own right!), we're going to take a look at the essential processes and steps that guarantee readable, usable results in the shortest amount of time. If you want to produce even more effective scores, though, it's well worth exploring further to see just how effective the Score editor can be in this task.

Preparing Your MIDI Files

As the old adage goes – garbage in, garbage out – this is never more true than in the process of creating a score. Although it's easy enough to open any region in the Score editor so as to see a notation-based view of the music, it doesn't guarantee that the score is legible or playable. Arguably, the clearest example of this is strings. In your MIDI arrangement, it's highly likely that the entirety of the violins, violas, and cellos, for example, will be amassed to one generic string patch. Although this polyphonic approach makes sense for MIDI production, it isn't how players expect to read a score, as in reality, each instrumentalist will be expect to be presented with a single line, with the full score combining all the single lines onto a page.

The first step to prepare an effective score is to divide the music into a number of distinct lines (one for each player, or part) within your Logic arrangement. As we saw in the MIDI sequencing and instrument plug-ins chapter, there are a number of ways of doing this – from simply copying the part over and deleting notes in turn to features like Functions > Note Events > Voices to channels,

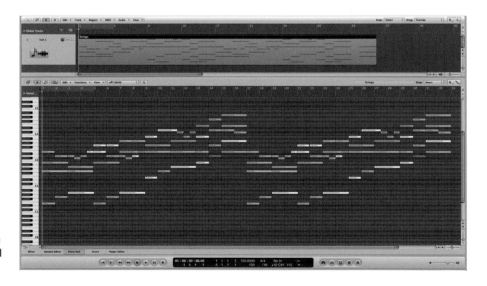

Figure 10.19
Here's a typical "massed" string part, with all the violins, violas, and cellos being triggered from the same general MIDI track.

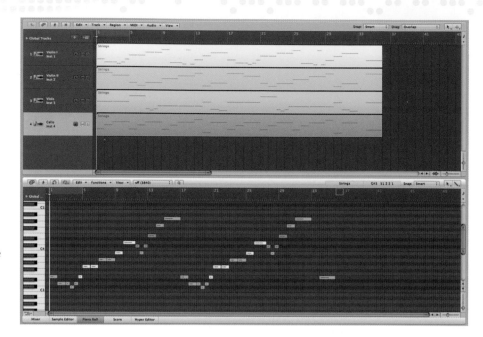

Figure 10.20
Creating a score from a series of split parts, each assigned to a different named track, will produce a more legible overall score, as well as facilitating the printing of individual parts.

followed by Split/Demix > Demix by Event Channel. Either way, you should end up with a series of regions for each part, each of which should be assigned to a unique named track for each instrument you want to appear in your score (like violin I, violin II, viola, and cello).

With the parts split, you might also want to glance through and double check any timing issues, as well as the precise duration of notes. For example, in a MIDI arrangement, it's easy to have notes finishing 1/16th before the end of a bar, but this can lead to some strange looking note durations in the score, alongside unwanted rests. For simplicity, it's also worth merging each part into a single region lasting the full length of the score, unless you want parts dropping in and out of the finished notation.

Opening the Score Editor

With the MIDI data prepared, you can now go and open the required parts in the Score editor, by drag-enclosing the required instruments and opening the Score editor via the Window menu. Although, of course, you can also open the Score editor directly from the Arrange window, it probably makes most sense to have the complete screen dominated by the scoring features.

On opening the Score editor, you should have something on screen that approximates music, although as with the raw MIDI data, this will need to be

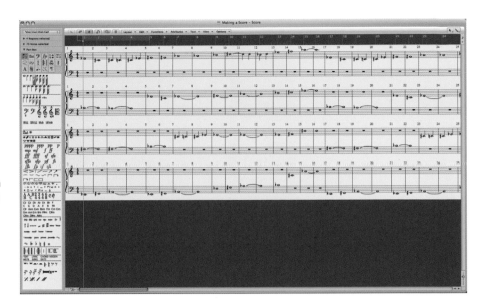

Figure 10.21 On opening the Score editor, you'll be presented with a basic score, although this will need to be modified to make readable.

tweaked to make the notation as legible as possible. For example, most of the instruments will probably default to a piano stave (with both a treble clef and bass clef on the same line!), and it's probably the case that the music simply runs across the full width of the screen, rather than being displayed on a page-by-page basis.

The first change, therefore, should be to turn the Score editor from its standard view into a page view, via View > Page View. Next comes the task of setting the staves. In keeping with the workflow on the Arrange window, you'll find that each line of score has its own set of score parameters, available via the Inspector to the left-hand side of the Score editor window (note that this might be minimized at first or even with the Inspector itself hidden). By clicking on each line of the score, therefore, you should be able to select an appropriate style for each part – maybe selecting treble for the violins or the bass clef for a cello part. Note that for any transposing instruments (like Horns in F), Logic will automatically transpose the "scored" version of the instrument, although the "played" version remains untouched.

Adding Expression and Score Markings

To add interest to the score, it is vitally important to add expression markings; otherwise you end up with a sterile and lifeless recording. To help with this, the Score editor provides a part box, which allows you to place a series of expression markings onto your score. These could be as simple as pp, mf, or ff dynamic markings, right up to trills, or crescendos and diminuendos. What you

Figure 10.22
Adding expression markings, and so on, will make your score both clearer and more expressive.

should notice is that the score stylings are attached to a particular part or line in question (the corresponding part will be highlighted blue when you drop the object) so that even when we move back down to part level (rather than the full score) these markings will be carried with the part.

Alongside expression marks, the part box also allows you to set the key of the piece (drag this to the first bar of the score and then any subsequent bar should a key change occur), as well as text-based objects like the song name.

Shrinking to Fit: Score Settings and Score Sets

Changing the broader qualities and appearance of the score can be done via the score tab of the project's settings, also assessable via the Layout menu. Although most of these settings (margin spacing and so on) will be best left in their default state, there are a number of useful functions. For example, the Number and Names tab allows you to change how instrument names are displayed, using either their Full Names throughout the entire score, for example, or an abbreviated Short Name for pages after the first page.

Another part of the full score's appearance is the percentage scaling of the staves on the page. For example, on a full score, it's often important to see as much of the music as possible, so as to avoid excessive page turning, as well

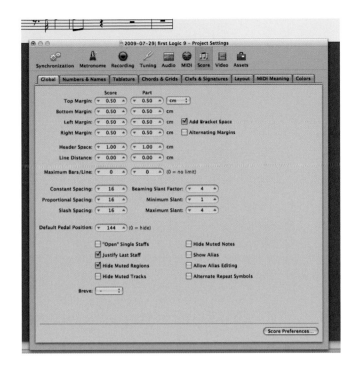

Figure 10.23 The Score tab under the project settings allows you to define some macro properties about how the score is laid out and printed.

visualizing the entirety of the arrangement at any point in time. This is of particular relevance to a full orchestral score, where you might need 30–40 different staves on the page at any one time.

Technically, when we selected our particular group of regions, we created what is known as a score set. As the name suggests, a score set assembles a group of instruments that will appear together on the score at the same time, theoretically letting you omit certain tracks on the Arrange pane from being displayed in the finished score. In addition to this, you can also define a percentage scale of a score set, allowing you to shrink or expand the score's size accordingly. Go to Layout > Score Sets... to access the Score Sets Window, which you'll need to ensure has its "local" Inspector open in the window so as to access the Scale [%] parameter.

Printing the Parts

To print off individual parts from the score, try double-clicking on the stave in question. This should drop you down one level (the equivalent of going into a folder in the Arrange pane), allowing you to see a single part with its associated score markings. So as to make the part clear and legible, you might want to consider raising the Scale [%] parameter as part of the Score Sets controls.

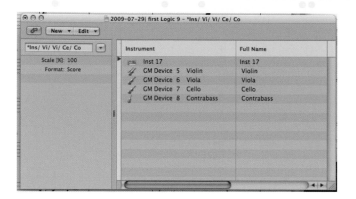

Figure 10.24 Use the Score Sets Window to alter the overall percentage scaling of the score.

Remember, though, that with both parts and the full score, you have the option to print to a PDF (via the OS) so as to create a saveable "hard" copy of the score for future reference, printing, or indeed, as an e-mail attachment for your music copyist.

10.6 Surround Sound in Logic

Surround sound, while synonymous with the film industry offering that extra dimension of realism, is now becoming popular in some circles for music production with many artists' back catalogs being mixed for the medium. Whether working to picture or in music production, you'll inevitably find yourself being expected to create a surround sound mix at some point. Fortunately, Logic has a great deal of expertise in offering surround sound with the inclusion of surround sound panners on each channel and a 5.1 Reverb plug-in amongst other features.

To work with surround sound, in this example 5.1, it is important to ensure that you have the speakers you require to monitor each output. You'll need your main stereo pair of monitors, plus a center speaker, two surrounds (left and right) plus a subwoofer ideally. Connecting these up will usually mean a direct connection from your audio interface or desk to the amplifiers or powered monitors. The connections you make are governed by the Surround Preferences (Logic Pro > Preferences > Audio > I/O Assignments > Output). Within this pane are three tabs relating to the Output, Bounce Extensions, and Input.

The input arrangement should follow logically the output assignments and is used for making 5.1 recordings. The bounce extensions tab simply refers to the filename extension which is important when preparing for mastering. Concentrating on the Output tab, you can select how Logic shows its outputs. Typically, this will default to Logic's own interpretation of 5.1. There are options for International Telecommunications Union's (ITU) and WG-4's output arrangements. The ITU's is standard for 5.1 surround sound for most professionals.

However, the WG-4 standard, which can be selected by the button below, is the choice of the DVD forum. At this stage, it is just important that the right outputs are connected to the correct speakers for accurate monitoring.

Getting Started

To get started with surround sound in Logic, you could either select a surround sound template from the chooser or alter your current stereo production to work in surround. Choosing Logic's own template offers 24 surround-ready channel strips all bussed to groups including a 5.1 incarnation of Space Designer. This is excellent for that 5.1 project you're about to start.

However, there will be times when you'll need to alter your session to output in surround. To do this, simply choose the channel strips you wish to pan in surround and then visit the output assignment button. The drop-down menu should now show the possible output assignments, the last of which is Surround. Click this and notice the master channel strip alter from stereo containing two peak meters to a surround fader with 5 meters.

Working with Surround is very similar operationally to that of stereo except panning is different. You will have noticed a circular panner on each channel. When double-clicked, this opens up the Surround Balancer pane that shows a good-sized panner showing a visual representation of the space with the five speakers facing inward. To pan a mono signal, simply drag around the ball within the space to inform Logic where you'd like it to come from. Logic will calculate how much signal should come from each speaker.

Figure 10.25
Selecting which audio outputs form your interface are sent to which speakers can be an important part of setting up your surround sound studio.

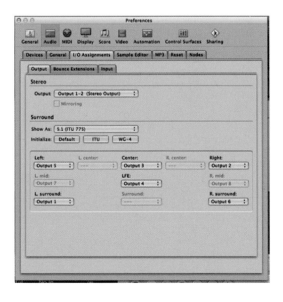

For stereo signals, Logic's Surround Balancer shows three balls. The "L" and "R" balls represent the two discrete mono signals that go to make up the stereo signal. These balls are linked and allow for you to consider how wide you want the stereo signal to go within the surround space. The third ball is the overall pan that will alter the direction the stereo image comes from. In the larger surround panner on p. 361, the stereo signal is shown biased to the right.

Other controls on the Surround Balancer include the center level and the low-frequency effect (LFE) level. The center level refers to the image created by the center speaker. As surround sound is derived from the film industry, the center

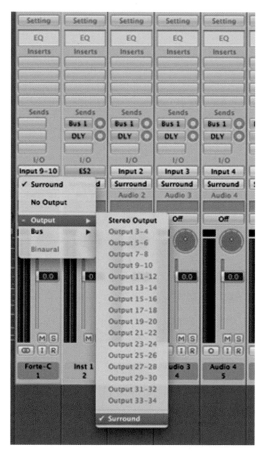

Figures 10.26 The Output Assign menu on each channel offers you the ability to choose any available physical mono or stereo outputs, in addition to a surround mode.

Figure 10.27 Clicking and holding on the button with a circle(s) in it will offer you the opportunity to quickly select the output mode of the selected strip.

speaker has traditionally been placed behind or around the projection screen and due to its focal point it contained all the dialog. As such, the music was usually omitted out of this speaker, leaving it to create the phantom center image it currently creates with stereo reproduction. With a center level control in Logic, it is possible to decide whether to allow any signal to come through the center speaker at all.

Panning stereo signals in surround can make the process of mixing somewhat complicated. With two speakers, it is clear where the stereo signals go and how they might be managed as they can only come from within the space provided between the two speakers.

Surround sound in its popular incarnation means that there are five speakers through which the stereo signal can be routed. Does this stereo signal simply go to the left- and right-hand speakers and faithfully represent itself, or is it reinterpreted for surround and maneuvered around the sound space? In the

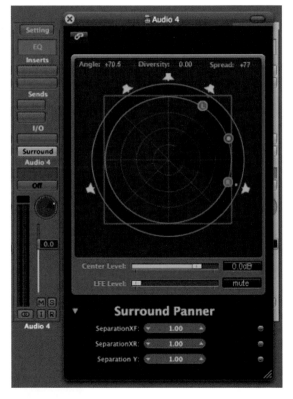

Figures 10.28 and 10.29 In this view, the panner can be seen managing a stereo signal. When panning stereo sources, there are extra controls to change the spread of the signals.

past, a stereo pair on a guitar might have stayed centralized to the center, but with a multichannel set up, it is possible to say rotate this 90° so that the left-hand channel could be reproduced from the right monitor and the original right channel from the rear right monitor, hence keeping the stereo image but shifting it in the sound space as shown in the example below.

Three further controls are available from the bottom of the Surround Balancer. To obtain access to these, simply click on the small triangle to the bottom left of the pane. Three sliders appear that offer you the ability to alter the separation of the panner. In the diagram above, the orange square represents the bounds of the possible separation of the panners. However, in some examples, it might be necessary to reduce this to reproduce a less wide image of the original. The three sliders correspond to the front left and right image (Separation XF), rear left and right image (Separation XR), and separation between front and rear (Separation Y).

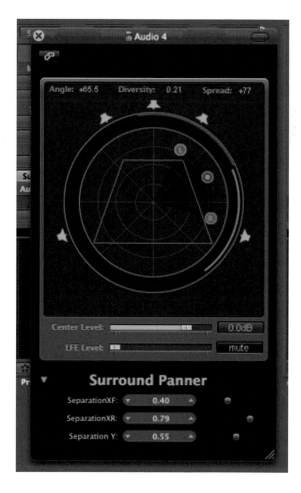

Figure 10.30 In this example, the separation has been reduced to limit the width of the image. Also, it is possible to spread circular lines around the circumference that indicate which speakers are receiving a signal from which channel.

The LFE level, sometimes also called the low-frequency effect, is the "0.1" of 5.1. This is what is commonly known in the industry as a subwoofer. The LFE level control is placed within each panner as it is often undesirable to send everything to the subwoofer, when the main monitors should handle most bass information perfectly well. This again has come from the film industry where the LFE is used to reproduce rumbles, explosions, and impacts. The 5.1 standard means that all the main five monitors should be full-frequency and should reproduce bass well for most applications. Hence, the LFE is included to add weight to the bass end. However, it is worth remembering when mixing that for many audio applications, including home cinema systems, the subwoofer is not included as an "effect" but as the only bass generating driver, leaving the satellites to consider only mid- to high-frequency ranges.

Knowledgebase ▽

Surround Sound Plug-Ins

Working in surround sound requires that you think slightly differently when employing effects such as reverberations. You could employ a stereo plug-in or you could choose to use two in tandem for front and rear if need be. Logic has fortunately thought of this and produced an intuitive set of surround-ready plug-ins to get you instantly working.

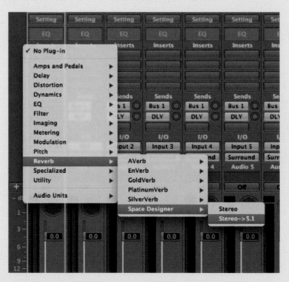

Figure 10.31 Choosing a surround plug-in can improve your 5.1 mix. Note how a mono track (or in this example a stereo track) will be processed with a surround sound plug-in thus changing its output to 5.1.

(Continued)

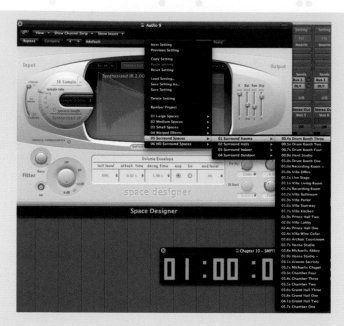

Figure 10.32 Space Designer comes with some impressive dedicated surround algorithms.

These dedicated plug-ins include the renowned Space Designer whose 5.1 algorithms are superb. Other 5.1 dedicated plug-ins include Delay Designer, Chorus, Tremelo, Flanger, Microphaser, and Modulation Delay.

The plug-in behaves in exactly the same way as the stereo version, but the response is returned in 5.1 as per the original impulse. To place a surround plug-in in your mix can be simply achieved by selecting it from the menu. Most 5.1 plug-ins can be applied to a mono track whose output will be converted to surround. However, these processes are best added to a 5.1 bus or output.

Other plug-ins do not require specific surround algorithms to operate with multichannel information. These will, therefore, work in "Multi Mono" mode for processing multichannel information. Plug-ins such as the compressor, which utilizes this mode, will show some banks using additional switches located above the main controls that are not found on its stereo or mono counterpart. The Configure switch allows you to configure which compressor bank is attached to which outputs. Therefore, it is possible to strap a compressor across the front speakers, another to the rears, and one dedicated to the LFE channel. Alternatively, you might wish to alter this so that the Center channel has a dedicated bank for use in film work where the dialog is of paramount importance.

Figures 10.33 and 10.34 Plug-ins that do not require specific surround algorithms work in Multi Mono mode such as the compressor. Note the selector switches above the main plug-in to select which compressor bank operates which output.

10.7 Delivery Formats

Bouncing to Stereo and Surround

At some point, it will be necessary to present a mixed output of your production. If you are working within Logic, then it will be necessary to do a bounce, and depending on the application, different choices will need to be made. One of these choices will be of course whether to produce a stereo or a surround bounce. Bouncing surround mixes is achieved in the same way as traditional stereo mixes.

Once your surround sound mix is balanced and ready, simply select the "Bnce" button from the master channel strip in Logic's Mixer. A new dialog button will emerge showing a number of options for editing. The first to note is the destination for the files once you have bounced them. Choosing the arrow to the top right will open up this dialog to show more of the file structure of the computer.

Moving on to the left-hand side, the Destination table can be seen, which allows you to configure the file type you wish to export to. For most applications, this

Figure 10.35
Bouncing to surround is as simple as clicking the "Bnce" button on the master fader strip. It is important to select the surround bounce checkbox and choose whether you wish the file to be interleaved or separated out.

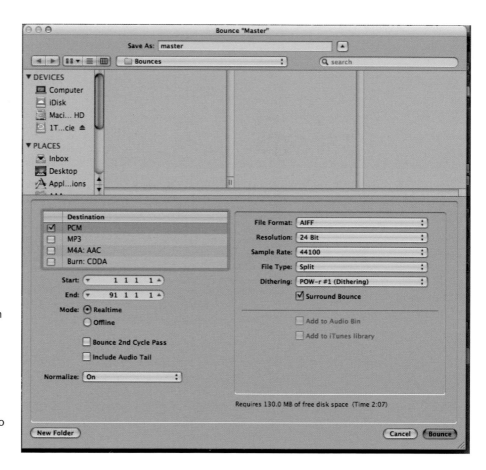

is likely to be PCM, which means unaltered or data compressed information, unlike the MP3 and M4A options below it. The DVD-A option is an excellent option for surround sound as it will create a disc that can be played in most DVD players connected to a surround sound system.

Below this are the range controls allowing you to specify the start and end points of your bounce. This might be useful in the instance that you might need to export each scene separately for a dubbing mixer. Additional features include whether you wish to monitor the bounce in real time or allow the computer to manage the event offline. You are also given the opportunity to automatically normalize your audio as it is bounced.

The right-hand side of the dialog box includes a number of drop-down menus that relate to the file destination type you wish to export. First up is the option to choose the file type. There are four choices here starting with Sound Designer II. The SDII format was created by Digidesign and has remained a standard for many years. Its benefit is that it is generally time-stamped meaning that it can be moved and can be recalled to its original position at any time. This can be extremely useful when working to film, for example, or when transferring session between DAWs which accept the SDII format. The second incarnation is the Audio Interchange File Format (AIFF), which is generally the common format used on the Apple Mac. This file format is not subject to any data compression and is generally an accepted file format for use between software and the industry. AIFF does not contain timestamp information. Next is the Wave format or .wav and is perhaps the most common of audio file formats for exporting to another computer for further mixing or mastering. Finally, in the menu is the Core Audio File, which is an all-encompassing format that contains a wide variety of different formats in one and is common for Apple Loops.

The next drop-down menu is labeled Resolution and relates to the bit depth you wish for your bounce. This can be set typically at 8, 16, or 24 bits. Following this is the sample rate selector where a wide variety can be chosen from for a wide variety of applications, which will be determined by your destination. The File Type box allows you to set whether you wish for the files to be interleaved together or split files for each channel.

Dithering is required in situations where a lower resolution bounce is required than the project's set resolution – for example, when a 24-bit recording needs to be dithered down to produce a 16-bit CD ready file. Dither is a method by which the waveform is more accurately encoded for 16-bit reproduction and has been discussed previously.

The next, and perhaps most important feature for delivery of a multichannel export, is the Surround Bounce checkbox. This needs to be checked to enable that each channel of the surround field is exported as a separate file. The extensions for which are changed in the Audio Preferences as described before.

Exporting to DVD

Logic now allows you to export a DVD-Audio ready disc directly from the bounce dialog. Within this window, select "Burn: DVD-A" and the right-hand pane should alter to show you the Mode. There are two choices here. DVD-A will produce a 5.1 mix for use in DVD-A players, while "CDDA" allows for a high-definition (HD) stereo bounce of 24 bits and a 192-kHz sample rate.

For working to DVD Video, surround sound files need to be compressed using Dolby Digital Professional (AC-3) format. To achieve this, we need to use Apple's Compressor software which is part of Logic Studio or similar encoder software – see Knowledgebase.

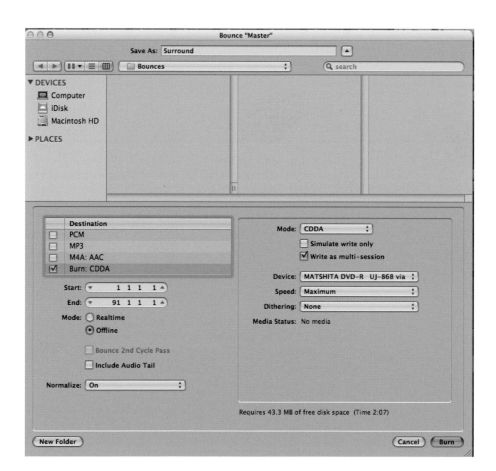

Figure 10.36
Burning a DVD-A or CCDA can be achieved straight from Logic's own Bounce window.

Walkthrough ▼

Creating AC3 Files Using Apple Compressor

For your new 5.1 mix to be read on a DVD-V disc, it will need to be compressed to Dolby's AC3 format. To achieve this using Logic Studio, you will need to export the six audio files (front left, center, front right, rear left, rear right, and LFE) using the "bnce" bounce feature. With six discreet audio files, it is now possible to compress for AC3.

To do this, make sure you've got your six audio files to hand. Quit Logic and search your applications folder for Compressor. Open this. You'll note Compressor looks quite different, but is a fantastic batch processing tool for all kinds of conversions and so on.

In Compressor, go to Job > New Job with Surround Sound Group.

A new dialog box will appear giving you options to place individual, discreet audio files in each speaker. You can add 5.1 or indeed an additional rear center channel "S" making this 6.1.

Also note the ability to add a movie here to compress the whole thing together for a DVD-V ready file.

Figure 10.37

(Continued)

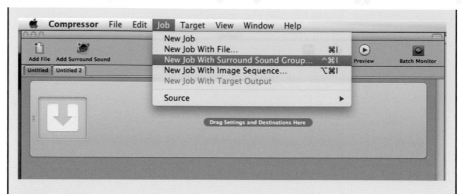

Figure 10.38

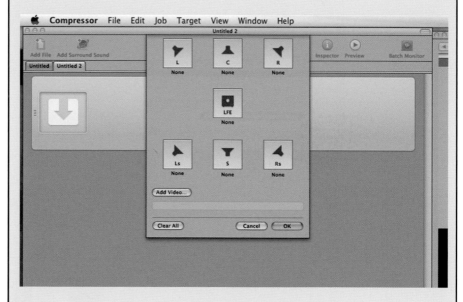

Figure 10.39

Once you've clicked OK, you'll be back on the main Compressor screen and it is at this point you need to specify what the output file format will be. To do this go to Window > Settings or simply click Cmd+3.

A new Setting dialog box will open that has two tabs: one for the Destinations of compressed file and one for the Settings. We're interested in the Settings at this point. There are a large number of file formats

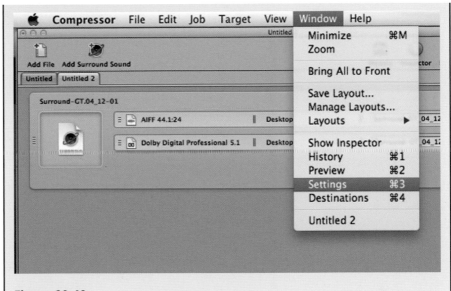

Figure 10.40

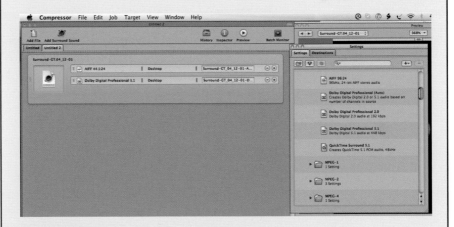

Figure 10.41

here to choose from, but seek out Formats > Audio > Dolby Digital Professional 5.1. Drag the Dolby Digital Professional 5.1 icon to the batch job in the main screen. Next press Submit on the bottom right to press your AC3 file.

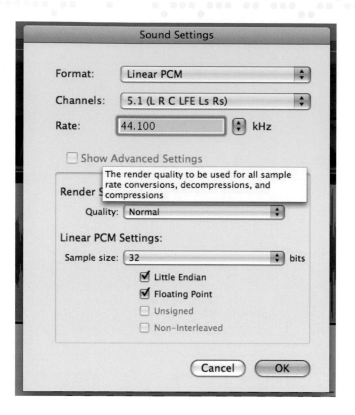

Figure 10.42
This pane allows for some choices for the nature of the exported audio.

First, we need to bounce the Surround Sound mix as split files, meaning that each stream (left, center, right, left surround, right surround, and LFE) is an audio file in its own right. This will then allow us to indicate in Compressor what files are for which output. Before bouncing, it might be worth checking the bounce extensions found in the audio preferences pane as discussed earlier.

Exporting Video with Audio (Dubbing)

Either during or at the end of the writing process for the film and television visual, you will be asked to send a copy with both the video and audio together into one file. This process is known in the industry as dubbing and is something that is normally handled by large postproduction houses, but for demo purposes, you can merge the two files together by selecting Export Audio to Movie from the movies menu (File > Export Audio to Movie...).

A new dialog box entitled Sound Setting will appear that allows you to choose the nature of the audio file. It is here that you decide upon the data compression format, sample rate, bit rate, and whether it should be stereo or mono. These settings are useful as a director or production company may wish to see your work in progress, and the best vehicle to deliver this to them is over the Internet via e-mail (if small) or File Transfer Protocol. It is, therefore, handy that the compression formats allow for a wide variety of standards to suit each and every client.

Once these settings are decided upon, click OK and choose the location for your final files. It is worth creating a folder at this point to ensure quick and tidy file management as there may be more than one file given the compression format you have chosen. Next click Save and wait while Logic compiles your video for you. Once completed, this file should be readable by Quicktime.

In This Chapter

11.1 Introduction 375

11.2 Templates 375

11.3 Screensets and Windows 377

11.4 Key Commands 382

11.5 The Environment 384

11.6 Input/Output Labels 400

11.7 Nodes and Distributed Audio Processing 401

Walkthrough

Adding New MIDI Devices to the Environment 393

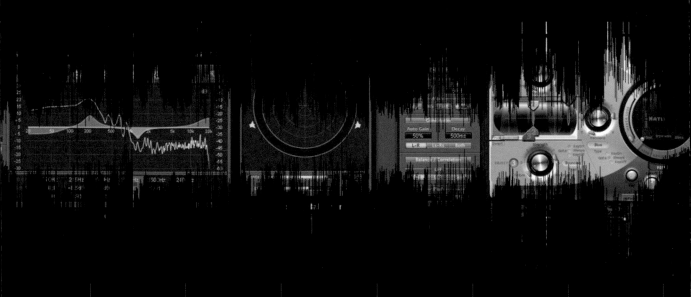

Optimizing Logic

11.1 Introduction

Working with any music software will expect you to operate within its structure and design. Logic is part of that pack in that offers a number of options to personalize the way in which you work with the application. One of the most exciting things about Logic is its flexibility and configurability to the way in which you need and like to work. Music production requires a fluid workflow that adapts to your way of making music. Doing this requires some initial thought and preparation, but it will improve not only the results but also the speed at which you work.

Over the years, Logic has developed an arsenal of features, tools, and techniques to help the workflow of productions make the best use of the computer's interface. In this chapter, we'll explore some of those features and how to tweak Logic to fit your workflow.

11.2 Templates

As you launch Logic, you cannot fail to notice the chooser that allows you to get started with one of the program's in-built templates. Many of these are very workable and are an excellent starting point. However, as you begin to work in a particular way, you'll want your settings just so, thus saving your time and allowing you to be more creative. It is, therefore, worth preparing a template of your own with all the connections, plug-ins, windows, and other features ready to go to work.

It is best to spend some time thinking about the way this template will look and function. Consider the way in which you use Logic before setting up the ultimate template. You may need to set up different templates for doing different projects in a wide variety of genres. To do this, load up the Empty Project template from the "Explore" collection in the Template Chooser. Alternatively, you may just wish to alter a template from the chooser, which suits the way

DOI: 10.1016/B978-0-240-52193-0.00012-5

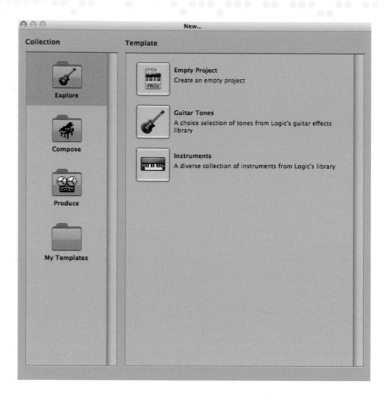

Figure 11.1
Logic's template chooser can save time when starting a project using a template of your own making.

you work with your personal samples already preloaded into EXS24. Add to this project your external MIDI devices that you use frequently and give them tracks, your preferred organization of Screensets (covered later), and then choose to save as a template (File > Save as Template…).

For example, if you do a lot of drum recording, with a guide bass, guitar, and vocals, it might be prudent to set up a template for this kind of recording, and perhaps, another template for working to picture. The benefits here are that all the tracks will be pre-labelled and, therefore, your audio files too when you hit record. This can be extended further to consider the mix, and if you find yourself using the same plug-ins and settings for your components of the drums, the template can have these preloaded as in the diagram opposite. It is perhaps sensible to put these into bypass mode (Alt + click) as this can save on DSP power. In this example, the drum outputs are all set to go to Bus 3, which in this case has been set up as a Drum Group or Stem.

Once you have created and saved your templates, they will appear in a new collection Folder within the Template Chooser called "My Templates." These should provide you with personalized and honed platforms upon which to move forward fast with your projects.

Figure 11.2
Creating a template for a recording such as this can save time in the long run. Each strip can be automatically set up for sends and some with plug-ins as required. Ready to press 'R' and record!

11.3 Screensets and Windows

Until Logic Pro 8, there were many individual windows to contend with. Each editor, mixer, list, and the Arrange window were all separate and individually controllable windows. That's hard to believe now with the redesigned Arrange window with its "access all areas" philosophy. Logic's solution to managing all those errant windows remains and is still very relevant in certain circumstances.

Working in Logic may require that a significant number of windows be opened on top of each other at a time. The most obvious examples would be an Arrange window and an editor open at the same time, or the Arrange window and Mixer together. Clearly, the current Arrange window can deal with most of these adequately. However, if you wish to open other windows, most applications would use the Command + " ` " (the one to the left of the "Z" key) shortcut to allow you to toggle through the open windows in the program one by one. However, given the need to access so many different combinations of the Arrange window, editors, mixers, etc., Logic adopts what it called Screensets to allow for quick and easy navigation.

Screensets are Logic's ability to change what windows are viewed on the screen, or screens, at any time (a little like Mac OSX Spaces). On the title bar of Logic to the right-hand side of the Screenset menu, you will notice a number. This indicates the screenset you have chosen, and pressing one of the numbers on the numeric keypad can change this, or you can click on the Screenset menu for more options. Within this menu, you have the opportunity to rename your Screensets, duplicate them, and delete them. The key additional features available within this menu are "Lock" and "Revert to Saved."

Lock allows you to set the way in which the screen looks so that it cannot be altered. A small bullet dot will appear to the left of the screenset number to indicate this. This can be really useful when you need to rely on your Screensets being just so! There will be times when the screenset changes you have made don't work for you and you wish to revert to your original decisions without losing the audio recording and editing you have done. To do this, choose the Revert to Saved option from the Screensets menu.

The window combinations can be carefully thought out to maximize both your speed and screen space. For example, most users will keep Screenset 1 as the default Arrange window. However, whatever combination of windows you choose for each Screenset is entirely up to you. One example some users use is to follow the window keyboard commands as a guide; see Window from the main Logic menu. Doing this means that you can get to navigate to the Screenset, which focuses on the task you need. For example, Screenset 8 could be set up to be the Environment, or the Audio Mixer within the Environment window, which we'll cover later in this chapter.

Screensets really come into their own with two monitors. As we've already mentioned, the new Arrange window is a very comprehensive space for achieving most things in Logic, although there will be particular times when you'll

Figure 11.3
The Screenset menu not only gives you options to edit and work with Screensets but also offers a clear indicator from the menu bar of which Screenset you're currently working in.

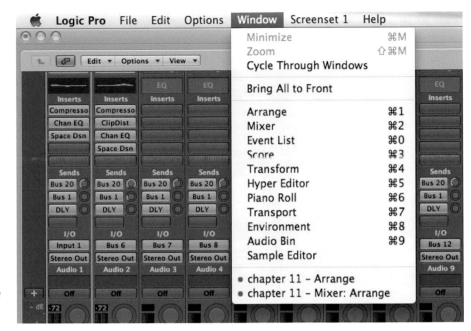

Figure 11.4
The Windows menu offers all the other views that can be spread across additional screens should you not wish to use the integral panes in the Arrange page.

want to see additional things, perhaps in larger windows or at the same time as something else. As such, Screensets could allow you, if you so desired, to manage the second monitor's content, while leaving the Arrange window on the main monitor. Additionally, Screensets can also permit you to choose which editor or list is shown within the Arrange window. For example, one Screenset could show the Sample Editor, while another shows the Markers list.

Should you wish to see the Mixer and the Arrange window as in the picture opposite, then this is absolutely fine. Although there is another more integrated way of bringing up the Mixer using the Environment window, Logic allows for the Environment window to be loaded up as a floating window by pressing Alt as you go to Window > Environment. Before you do this, open a normal Environment window by going to Window > Environment, and change its view to "Mixer" using the arrow pointing down in the top left-hand side of the screen.

With this new floating Mixer window, shown below, it is possible to quickly navigate each track on the Arrange window and the Mixer will follow as though the link control was on, making it an extremely flexible and powerful way of working efficiently whether on a small, wide, or double-screen setup. To enable this, you need to ensure that the link icon is enabled in the main Environment window before making it float. This might seem as duplication to the Inspector's expanded feature showing two Mixer elements, although you can, perhaps, be more dependent on the number of monitors you have. The alternative use for this is perhaps

Figures 11.5 and 11.6 Two windows can be made to sit side-by-side using Screensets as shown here, but it is also possible to create a floating Environment window, which can be linked to your actions on the Arrange area (below).

where you might disable the link and focus the floating Environment window (as a mixer) to the Master output. Now, directly from the Arrange window, we operate the focused track's Mixer element in addition to the master bus.

Figure 11.7
It is possible to import your project working preferences from another Logic File using the Import Settings feature.

Immediately, there are nine Screenset levels that can be accessed from the number keys on both the qwerty and numeric keypads. However, Logic allows you to access and make use of Screensets up to "99." To key in these Screeensets, you need to press Ctrl and type the number of the corresponding Screenset. So for 34, simply press Ctrl + 3 and then Ctrl + 4, on the numbers on the qwerty keyboard.

At times you may receive a file you have been asked to mix, which will have different Screensets and perhaps different key command settings. It is important that your workflow is not jeopardized and that you can work both efficiently and productively with your preferences. To import project settings from another Logic File, go to File > Project Settings > Import Project Settings.

Link Modes

When you create multi-windowed Screensets, it is often a good idea to use one of the link modes to connect them all together when editing. As we have mentioned earlier in the book, Link allows you to click on another region and its contents to be displayed in the accompanying editor. This is highly beneficial and ensures that you can work really fast while keeping all the windows you need open. If this feature did not exist you'd need to close the window and perhaps double-click on the new region each time. Laborious!

Link has two discrete modes. The first is Same Link Level (pink icon), which means that whatever item you click on, whether that be a region, or the note, the linked windows will follow the same level. Confusing isn't it? Well this can be useful if you have a number of editors open at any one time. A good way to see this working is by opening a separate Events List window (Cmd + 0) and, within the Arrange area, selecting a MIDI region. The Events List should show a list of the regions. However, if you press "P" to open up the Piano Roll Editor and then select on the MIDI notes, the Events List will change to the same level or class of data, in this case a MIDI note within all the notes in the region.

Content Link (Yellow Icon) differs in that you always see the content of the region. So in the example above, if we clicked on the region in the Arrange area, the Events List represented a list of the regions, but if in Content Link Mode we selected a MIDI note in the Piano Roll, then the MIDI note would be shown in the Events List. With Content Link, we can simply click on the MIDI region in the Arrange area to immediately see its contents in the Events List, therefore seeing the content level without having to select it separately in the Piano Roll.

These differences are ever so subtle, and frankly, it is one of those things that if the windows are not linking as you need them to, then simply Ctrl + click the link icon, change the mode, and see if that works the way you want it to. Nevertheless, the Link mode is a key feature of Logic as it does ensure that Screensets can operate fluidly without the need to shut an editor down to then double-click an alternative region to look at its notes!

11.4 Key Commands

Keyboard shortcuts are a useful way of improving the workflow with whatever software you are using from a word processor to an Internet browser. Logic, of course, is no different, and as you'd expect, there are so many of them covering a vast number of the features. Throughout this book, we give you the default key commands that relate to what we have covered. Most of these commands can be learnt by noting the writing to the right of the command in one of the menus, but the whole range can be viewed in their own dedicated Key Commands viewer (Logic Pro > Preferences > Key Commands... or Alt + K).

Searching for Key Commands is easy by using the Spotlight-like search bar on the top right of the pane. In here, simply type in the command you wish to find the key command for and the possible results will appear under one of the main headings in the command list. Under each of these 17 headings, are all the key commands available for use with Logic. Not all of these commands will have a shortcut associated with them as there are simply too many. But as with

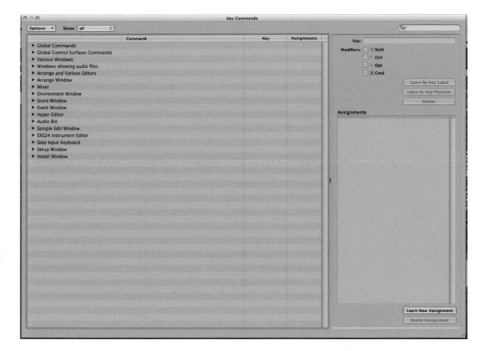

Figure 11.8
Key Commands are an important feature of any piece of software to increase productivity. Logic is no exception allowing you to carve out a key command for literally every feature.

all things in Logic, they too can be optimized or altered to suit your working needs.

Once you have selected the instruction you wish to create a key command for, select it and choose to either "Learn by Key Label" or "Learn by Key Position" followed by the key combination you wish to use. There are subtle differences between these buttons. Logic allows you to discern between using the numerical keypad and using the numbers on the main part of the keyboard. In this way, you could choose to assign two different commands to number 9, one would be on the main keyboard and the other on the numerical keypad. Therefore, learning by Key Label means that using the number 1 will call up the same key command whether it is on the numerical keypad or above the qwerty. However, the Key Position option will allow you to discern between the two differing positions, thus giving you further key command possibilities.

There will, perhaps, come a time when you will develop your own key commands and see your productivity suit the way in which you work. Your templates will already preload your key commands with your work if you saved them, but there may come a time when you have to work on an imported project such as a mix that someone else put together. In this instance, it is really important that you save a set of your personalized key commands.

There will be times when you may need to impose your special and personalized changes to another's Logic session and will need to have your key commands ready. To do this, we need to export our key commands. To do this, simply press Alt + K from anywhere in Logic to bring up the Key Commands dialog. Within the Options menu is a command called Export Key Commands which will allow you to save your key commands. These can then easily be exported into another copy of Logic using the Import Key Commands instruction in the same menu. These key commands can also be saved to the clipboard for use as a reference, or for revision. It may even be worth doing this while you're learning to use these commands.

If, however, you have learnt Logic's default Key Command set and import another person's project, you might be at a loss as to why so many of them did not work. In this case, it would be sensible to choose the option of "Initialize All Key Commands," which will reset the commands to the factory standard.

11.5 The Environment

Logic's Environment window has an interesting reputation to the non-Logic using world. It has always held an almost mythical-like status, where many users have not wished to delve for the fear of it would mess their whole project up. Although the Environment is perhaps less used these days now that MIDI is not the main method of music communication between the application and the sound sources, it is still an important tool that allows the user to experiment and perhaps optimize his or her creative output. The Environment window is thus the route for information into and out of Logic and as such is something that is worth considering especially when working with MIDI. It is here that the virtual representation of your studio can be created, managed, and manipulated. In this section, we'll give a quick overview of its core functions and its possibilities. Where you take it will be up to your creativity.

Layers and Windows

The Environment window itself could be simply looked upon as an area to organize your MIDI inputs and outputs and how they all connect together. However, there are some more angles, or layers, to it which we'll explore a little here such as its Audio Mixer and Global Object layers. In essence, the Environment should show all the connections to and from Logic, both physical and internal.

The first layer in the list is the All Objects list, which shows you all the connections to Logic currently established. This includes MIDI, Audio, and internal connections such as ReWire. The next layer is Global Objects that allows you to specify any objects that are common to each layer. The Click and Ports layer refers to how the click track (metronome) is managed within the MIDI environment, and how the Ports connect to each MIDI input, whether that be physical through an interface (port) or the Caps Lock Keyboard. The MIDI Instr., or MIDI

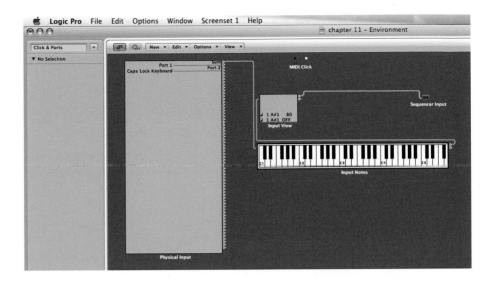

Figures 11.13 and 11.14 Using the Environment to create unique and interesting creative aspects to your work is easy. In the examples above, we have placed an Arpeggiator across any MIDI input to Logic.

Setting Up MIDI Instruments in the Environment

The MIDI Instruments Layer is an important one as it allows you to specify the MIDI environment and the synths and devices you have attached to your system. For example, if you bring in a new MIDI synth to your setup, you could simply just tell Logic there is a new instrument connected and this will allow you to select it from the track assignments, which is fine, but there is so much more that can be done.

Creating a new instrument within the MIDI Instruments layer will enable a new instrument to be located within the Reassign Track Object menu meaning that

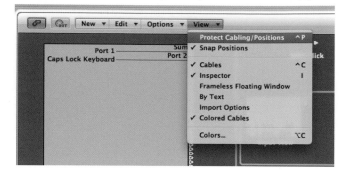

Figure 11.15
It is easy to view, hide, or protect the Environment's cables by going to View > Cables in the Environment.

you can directly select the device and its channel from the Arrange window. To do this, we need to create a new Environment Object, and there are two main alternatives for MIDI instruments. The first is to create a new "Instrument" (New > Instrument). This means that this output will go to one MIDI channel from any port. There are some classic examples of MIDI instruments which are not multitimbral such as the Waldorf Pulse, which is a single MIDI-channelled monosynth.

For larger multitimbral synthesizers, we'll need to create a "Multi-Instrument." This can allow us to specify quite a lot about the device in question. As you select this, you're confronted by a box with 16 small buttons. These will all be crossed off. This is essentially to save your time as not all multitimbral synthesizers are able to use all 16 MIDI channels. Therefore, you can specify which outputs you wish to use. Another reason for limiting the output of certain channels might be because you may wish to use a channel from the same physical output, if your MIDI outputs are limited, to send to a monosynth such as the pulse on channel 16.

After creating an instrument object, you will need to connect it to an output port. To do this, simply select the instrument in question and visit the Inspector to the left of the Environment window. Usually, directly under the Icon checkbox is a drop-down menu which lists all the available physical ports. Simply select the physical MIDI port you have connected the device to and you can use the instrument straight away from the Arrange window. This is known in Logic speak as a direct output assignment.

Once you have set the device using the multi-instrument object, you can double-click the object to reveal the Multi-Instrument Window. This immediately shows the instrument names for a General MIDI device (GM Device) that relate to a program change that can be sent by the sequencer. The idea is that you can relabel the instruments within Logic as they relate to their corresponding program change number. Once this has been done, you can choose the instrument name when selecting the Track Assignment in the Arrange window. Although this might seem like a long-winded exercise for all

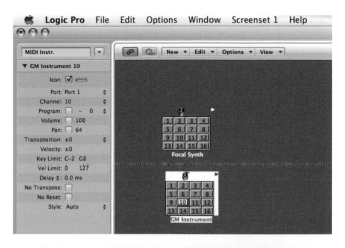

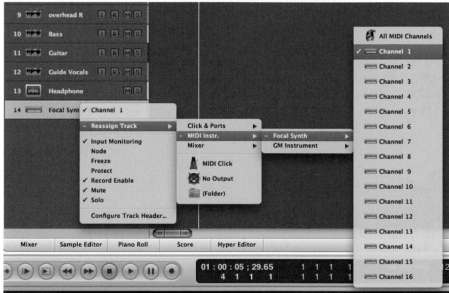

Figures 11.16 and 11.17
Creating a new Multi-Instrument in the MIDI Instruments layer will allow you to create a new stream of MIDI information intended for this new device.

your esoteric synths, the operation would only need to be completed once. That level of recall from the Arrange window is an excellent feature and once done can speed up your workflow without leaving your mix position.

The Mapped Instrument Object can be useful when you wish to convert one type of MIDI input data to another. For example, you may have drum kit loaded in an external device, which does not follow the General MIDI drum convention. In that instance, it would be useful to call up a Mapped Instrument Object which can be double-clicked to open a chart which allows one note to be mapped to another in real time.

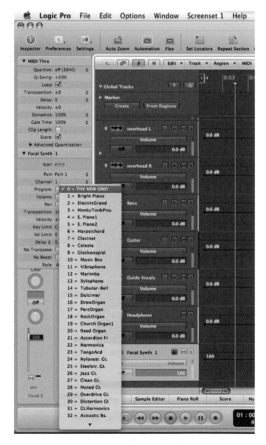

Figures 11.18 and 11.19
The Multi-Instrument window allows you to label the instruments or patches to their corresponding program change. This can be then selected from the Program Change menu in the Inspector.

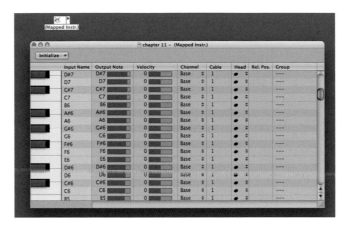

Figure 11.20 The Mapped Instrument Object converts one set of MIDI notes into another and is primarily useful for mapping drum kit sounds.

Walkthrough ▼

Adding New MIDI Devices to the Environment

Step 1:

Figure 11.21

Adding a new MIDI instrument to the Environment will allow you to select it as a distinct item from the Reassign Track Objects menu. First, open up the Environment window, which is Cmd + 8 or Windows > Environment. With the Environment open, select the layer selection menu to the top left.

(Continued)

track list on the Arrange area. This second method will only work, provided you have enabled the "Icon" check box which relates to the mixermap's Monitor Object.

Provided all the objects are cabled correctly, your newly created mixermap should offer real flexibility to control devices directly from your Mix position. The open-ended nature of the Environment allows for many more possibilities to control devices within your studio.

11.6 Input/Output Labels

When working within a large studio environment that brings together a fairly big analogue setup with many processors and a mixing console, there may be simply too many connections for you to remember which specific input relates to your prized Fairchild unit. Like with so many things within Logic, this can of course be labelled up on some physical inputs. To do this, select Options > Audio > I/O Labels... A new window opens showing all the physical inputs and auxiliaries within the system.

Figure 11.29 The Input/Output (I/O) Labels within Logic can be named to improve the workflow of your session. Naming the key inputs and outputs and auxiliaries makes things quicker.

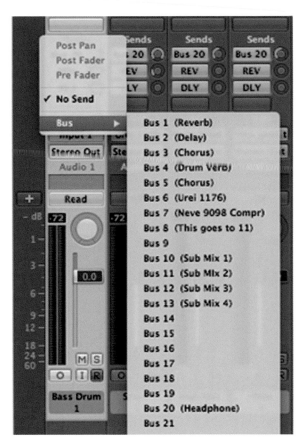

Figure 11.30 The changes made in the Input/Output (I/O) Label dialog box makes all the difference with respect to quickly recognizing busses in the Mixer.

To change the names of connections, click either in the Long Name or in the Short Name boxes. The Long Name is your description and should be whittled down to a smaller amount of characters for the Short Name which will be used more frequently when accessing drop-down menus in the Mixer.

11.7 Nodes and Distributed Audio Processing

In Chapter 3, we briefly referred to distributed audio processing and the notion of Nodes. Here, we discuss how to set them up, get the best out of them, and what to do when you want to take the project away without the additional power of the node.

Figure 11.31 The nodes can be easily set up using the Nodes pane within the Audio Preferences.

Connecting Nodes to your Logic setup is very straightforward. This is achieved by visiting the Audio Preferences pane and selecting the Nodes tab. There is a small dialog box, which allows you to "Enable Logic Nodes." With this option checked, Logic scans for computers connected to the host which have the Logic Node application running. It is not necessary to have Logic Pro installed, only the Logic Node program loaded on the slaved machine. Simply choose the computer you wish to act as the node and then, hey presto, this should allow for communication between Macs.

To make use of the node facility, you need to select which aspects of your production are handled externally by the node, or internally by the host computer. This selection is made possible through Track Node Checkboxes which can be revealed by visiting View > Configure Track Header.

The dialog box (below) appears offering you to customize the way in which you see the track's header, which is the area to the left of the arrangement and immediately to the right of the Inspector. Within the Configure Track Header dialog box, the options to change the buttons on offer are selectable. Some

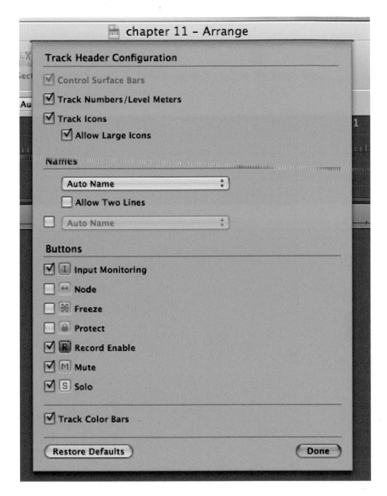

Figure 11.32 The Track Header Preferences allow for many changes including the Node Button asking whether or not a particular channel is to be managed by a Node.

useful buttons lurk here such as the Freeze Track icon and, in this instance, the Node button.

A good way of keeping an eye on how your processing is distributed is to call up the Load Meters (CPU/HD) which now reside within the transport bar. To turn these on and off, you'll need to Ctrl + Click the transport bar and then choose Customize Transport Bar... . Within here you'll be able to select Load Meters (CPU/HD) under the Displays list.

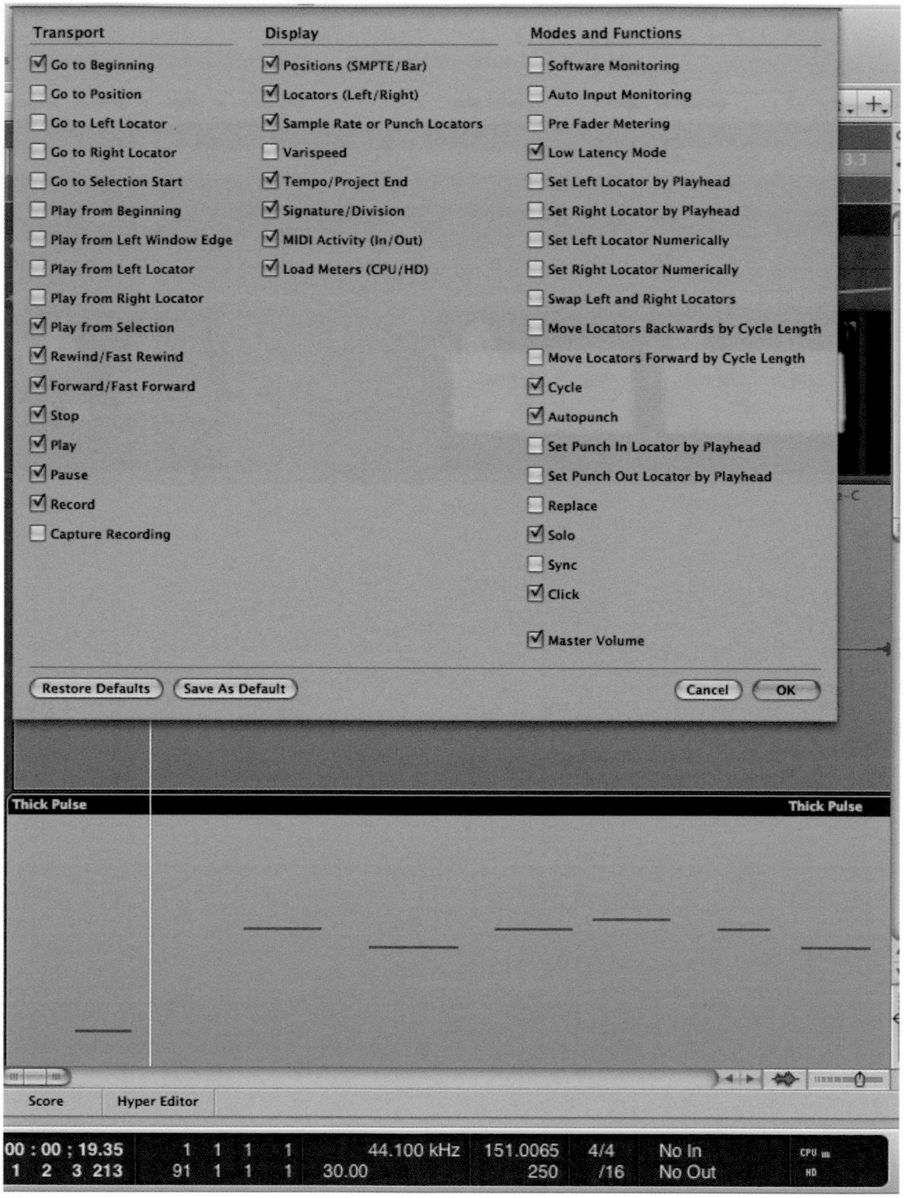

Transport	Display	Modes and Functions
☑ Go to Beginning	☑ Positions (SMPTE/Bar)	☐ Software Monitoring
☐ Go to Position	☑ Locators (Left/Right)	☐ Auto Input Monitoring
☐ Go to Left Locator	☑ Sample Rate or Punch Locators	☐ Pre Fader Metering
☐ Go to Right Locator	☐ Varispeed	☑ Low Latency Mode
☐ Go to Selection Start	☑ Tempo/Project End	☐ Set Left Locator by Playhead
☐ Play from Beginning	☑ Signature/Division	☐ Set Right Locator by Playhead
☐ Play from Left Window Edge	☑ MIDI Activity (In/Out)	☐ Set Left Locator Numerically
☐ Play from Left Locator	☑ Load Meters (CPU/HD)	☐ Set Right Locator Numerically
☐ Play from Right Locator		☐ Swap Left and Right Locators
☑ Play from Selection		☐ Move Locators Backwards by Cycle Length
☑ Rewind/Fast Rewind		☐ Move Locators Forward by Cycle Length
☑ Forward/Fast Forward		☑ Cycle
☑ Stop		☑ Autopunch
☑ Play		☐ Set Punch In Locator by Playhead
☑ Pause		☐ Set Punch Out Locator by Playhead
☑ Record		☐ Replace
☐ Capture Recording		☑ Solo
		☐ Sync
		☑ Click
		☑ Master Volume

(Restore Defaults) (Save As Default)   (Cancel) (OK)

Thick Pulse   **Thick Pulse**

Score	Hyper Editor

| 00 : 00 ; 19.35 | 1 | 1 | 1 | 1 | | 44.100 kHz | 151.0065 | 4/4 | No In | CPU ▦ |
| 1 | 2 | 3 | 213 | 91 | 1 | 1 | 1 | 30.00 | 250 | /16 | No Out | HD |

Figure 11.33 Keeping an eye on the computer load can be turned on or off using the Transport Bar customizations.

Within the Load Meters, the CPU bar has perhaps more than one lane in it, which represents the number of processors within your Mac, whilst the Disk I/O shows the data through-put to your hard drive. As a node is added it expands to show the node's CPU load, also allowing you to make a definite assessment of the project's DSP distribution.

Through employing many of these features and thus tailoring the software's behavior in the studio, Logic continues to be the trusted production tool for many professionals.

Subject Index

A

AC3 files, creating, 366–369
Advanced Quantizing options, 166–168
All Objects layer, 384
Alternatives in global tracks, 341
Amp Designer, 82–83, 261–262,
    263, 306
Amplifier, 204, 212
    envelope, 230
Amplitude envelope in sculpture, 230
Analog control of oscillators, 206
Anchor points, 127
Apogee, 29, 30, 32–35
Apple compressor, creating AC3 files
    using, 366–369
Apple Loops, 55, 73–77
    Column view, 74
    format of, 73
    importing, 77
    Music view, 74
    Sound Effects view, 74, 76
Arpeggiator, 388
Arrange window, 12–13, 377
    contextual menus in, 22
    editor areas in, 13–16
    inspector area in, 19–21
    list area in, 17–19
    local menus in, 22
    media area in, 17–18
    tools in, 21, 22
    transport bar in, 21
    zoom functions in, 23–25
Assets, 56–58
Audio Bin, 72–73
    for converting audio files, 139, 140
    for importing audio files and CDs,
      72–73
    tab *vs.* Audio Bin window, 137
    viewing, 139–141
Audio Bin window, 137
Audio channel strips, 260
Audio editing, 90
    cutting with Scissors Tool, 93–94
    fades and crossfades, 97–100

    Marquee Tool for, 95–97
    MIDI and, 7
    and processing, 322
    resizing, 93–97
    snapping, 90–93
    topping-and-tailing, 95
Audio files
    and CDs, 72–73
    management, 138–139
      copy/converting files, 139
    vs. MIDI files and instruments, 76
Audio hardware, 9
Audio Interchange File Format
    (AIFF), 364
Audio interfaces
    FireWire based, 32
    USB based, 28–30
Audio mastering, 314
    and CD burning, 330–331
Audio Mixer. *See also* **Mixer**
    input monitor in, 61–62
    mute in, 63
    record arm in, 63
    solo in, 63
Audio preferences, 40–42
Audio recording, 11–12
    in cycle mode, 67
Audio regions, 85
    arrangement, 86–87
    copies, 88–89
    Flex Time editing, 108
    groove templates between, 116–117
    loops, 89–90
    resizing, 126, 127
    separating, based on transient
      markers, 117–118
Audio spotting, 346–348
Audio tracks, 59, 146, 156
Auto map, 242
Automation
    basics of, 289
    curve tool, 298, 299
    menu, 300–301
    modes, 292–295

Automation (*Continued*)
  Latch, 294
  Read, 293–294
  Touch, 293
  Write, 292
  track-based *vs.* region-based, 290–292
  viewing and editing, 295–298
Autopunch, 68–69
Aux channel strips, 264–265, 281, 283–285

B
Band-pass (BP) filter, 211
Band-reject (BR) filter, 211
Bar ruler, 64
Basis tempo, 338
Beat mapping, 134–137
Big Bar display in global tracks, 346
Bit rates in digital audio, 41
Blend parameter, 217, 218
Bounce in place feature, 229
Bouncing and burning, 324–327
BPM Counter plug-in, 135
Buffer range process, 41
Bus processing, 283, 284

C
Cache, 338
Cakewalk UA25EX, 29
CD burning, 310, 317, 330–331
CD mastering. *See* Mastering
Channel EQ, 280–281
Channel faders, clipping, 270–271
Channel Strip Settings, 150–151
Channel strips, 260
  audio, 260
  aux, 264–265, 283–285
  in inspector area, 21
  instrument, 260–261
  master, 268
  MIDI, 267–268
  settings, 288–289
  stereo output, 265
Chord Memorizer Object, 395
Click and Ports layer, 384
Comping, 102
Component modeling, sculpture, 222
Compression, 276–278, 287
  circuit types, 302–303
  parallel, 287–288
  settings, 303–305
Compressor, 277–278

Comps, 103
  creating, 102–103
  flattening, 105–107
  time slipping, 104–105
  unpacking, 107
Condenser microphones, 35, 36
Content Link mode, 382
Contextual menus in Arrange window, 22
Contiguous zones, 243
Control surfaces, 43–45
  controller assignments in, 45
  Mackie's Control Universal, 43, 44, 45
  setting up, 44
Controller keyboards, 30–32
Convolution reverb, 284–285
Core audio, 40
Core audio file, 365
Correlation Meter, 275
CPU resources, saving, 118, 299–300
CPU-light plug-ins
  ES E, 165
  ES M, 165
  ES P, 166
Crossfades, 97–99, 326
Cue mix. *See* Headphone mixes
Cutoff parameter, 211
Cutting regions with Scissors Tool, 93–94
Cycle mode
  audio recording in, 67
  MIDI recording in, 158–160

D
DAE. *See* Digidesign audio engine
DeEsser, 297, 298
Delay plug-ins, 294–295
  sample, 295
  stereo, 295
  tape, 294, 295
Delivery formats of surround sound
  bouncing to stereo and surround, 363–365
  exporting to DVD, 365
  exporting video with audio, 369
Denoiser, 325
Digi CoreAudio Manager, 49
Digidesign audio engine (DAE), 6, 34, 49
Digidesign hardware
  integration with logic, 48–51
  setting up, 50
Direct TDM (DTDM), 49
Disc description protocol (DDP), 332
Distortion effects, 261–264

Distributed audio processing, 45–46, 401–405
Dithering, 365
 in mastering, 313, 324–327
Drag mode, 99, 100
Drive parameter, 218
Drum replacement, 121–123
Drums option, 243
Dubbing, 369
Duet devices, 30, 33, 34
Duplicate tracks, 70
Dynamic microphones, 36, 37

E
Edit groups, 119–121
Edit menu functions, 183–186
 Invert Selection, 183, 184, 185
 Select Equal Events, 186
 Select Equal Subpositions, 183, 185
Editing EXS24 instruments, 250–254
 filter activation, 251–252
 instrument refinement, 251
 modulation, 252–253
 saving instrument settings, 253–254
Editing fades in mastering, 314–316
EFM1 synthesizers, 201
 and frequency modulation synthesis, 216–222
 modulation on, 220–222
 refining patches, 222
Electret condenser microphones, 36, 37
Ensemble, 33, 34
Envelopes in ES2, 205, 212, 213
Environment, 384
 cables on, 387–388, 390
 external control with, 396–400
 layers in, 384–386
 Live MIDI in, 395–396
 MIDI instruments setting in, 389–391
 mixermap, 396–400
 objects in, 386–387
 window, 379, 384
Equalization in mixing, 276–278
Eraser Tool, 87
ES E, 165
ES M, 165
ES P, 166
ES2 synthesizers, 200, 202–205
 amplifier in, 204, 212
 envelopes in, 205, 212, 213
 filter routing system, 217–218
 filters in, 203, 210–215

global parameters in, 216
LFOs in, 204, 205, 215, 216
modulation matrix in, 204–205, 213–215
oscillators in, 202–203, 205–210
EVB3, 154–155
EVD6, 160–161
Event List editor, 193–194
 filtering events, 195
 modifying and adding events, 195–196
Events List Tab, 340
EVP88, 149–150
Export Key commands, 384
EXS24, 237–238
 data management, 196–197
 importing samples, 242–243
 instrument creation, 238–239, 242–243
 instrument editor, 239–240
 instruments editing. *See* Editing EXS24 instruments
 interface, 171–172
 mapping options in, 249–250
 multioutput versions of, 188
 REX files importing in, 240, 241
 virtual memory and disk streaming, 176–177
 working with, 248–250
External MIDI instruments
 adjusting delay settings in, 157–158
 monitoring, 155–157
 port settings, 153
 working with, 152–158
External MIDI tracks, 59

F
Faderport, 44
Fades
 creating by
  changing drag mode, 99, 100
  Crossfade Tool, 97–99
  fade parameter, 100
 editing, in mastering, 314–316
Filters in ES2, 203, 210–215
 cutoff and resonance, 211–212
 routing system, 217–218
FireWire devices, 28, 32–34, 42
Flex markers
 audio events flexing by, 111–112
 deleting, 114, 115
 notes flexing by, 112–114

Flex Time
editing, 108
modes, 108–111
monophonic, 109
polyphonic, 110
rhythmic, 109
slicing, 109
speed flex, 110–111
tempophone, 110
quantizing audio with, 115–117
view, 108–111
Folder
contents viewing, 142–143
and mixer, 273–275
packing, 141–142
Follow Tempo, 131
Freeze function, 300
Freezing track, 118
Frequency modulation (FM), synthesis of,
216–222
Functions menu, 183, 184
Note Force Legato, 186, 187
for note modifications, 186–187
Note Overlap Correction, 186
Select Highest Notes, 186–187
Select Lowest Notes, 186–187
Voices to Channels, 187

G
Gain plug-in, 274
Global Objects layer, 384
Global parameters in ES2, 216
Global tracks, 339–348
alternatives in, 341
Big Bar display in, 346
detect scene changes in, 341–342
markers in, 339–341
signature changes in, 345
tempo changes in, 342–345
Goniometer, 275
Group setting for mixing, 285–287
Guitar Amp Pro, 261, 263

H
Hard drives, 42–43
LaCie d2, 43
Headphone mixes, 77–79
Headphones, 38–39,
DT100, 39
Hiding tracks, 143
High-pass filter (HPF), 211
History of Logic Pro 9, 4–5
Hyper Draw in Arrange area, 182

Hyper Draw in Piano Roll editor. *See also*
Piano Roll editor
filter cutoff in, 181
velocity in, 180
volume in, 181
Hyper editor, 16, 17, 190–193
editing drums using, 193
Hyper sets addition, 192–193
MIDI controller data in, 190–191

I
Input monitor in Audio Mixer, 61–62
Input/output labels, 400–401
Inspector area in Arrange window, 19–21
Inspector's region parameters, 87, 88
Instrument channel strips, 260–261
Instrument Parameters, 151, 152
Instrument tracks, 59
assigning virtual instrument to, 147
creating, 146
Instruments icon, 151, 153
iTunes, 310

J
Jitter, 233, 234

K
Key commands, 382–384
Key Limit feature, 152
Key range, 244
Keyboards, USB controller, 30–32
Klopfgeist, 66

L
Latch, automation mode, 293
Latency, 51
accounting for processing, 81–82
and monitoring, 51–52
Layers, 384–386
LFOs. *See* Low-frequency oscillators
Library feature, instrument settings in,
147, 148
Limiter plug-in, 323
Linear-phase EQ, 315–316
Link modes, 381
Content Link, 382
Same Link level, 382
List area in Arrange window, 17–18, 19
Local menus in Arrange window, 22
Logic express *vs.* logic studio, 5–6
Logic studio
installing, 7–8
vs. logic express, 5–6

Logical advantages of Logic Pro 9, 6–9
Loops
    audio region, 89–90
    tempo and, 131–137
    in zone parameters, 247–248
Low-frequency effect (LFE) level, 360, 363
Low-frequency oscillators (LFOs), 204,
        205, 215, 216
Low-pass filter (LPF), 210

M

Mackie Onyx Satellite devices, 32
Mackie's Control Universal Pro, 13, 11, 15
Maestro, 33
Mapped Instrument Object, 391, 393
Markers, 339–341
Marquee Tool, 95–96, 97
Master channel strips, 268
Mastering, 309
    audio processing and editing, 322
    bounces to disk, 311–313
    and CD burning, 330–331
    different approaches, 309–311
    editing fades in, 314–316
    exporting and burning, 317–319
    tools, 324–325
    in WaveBurner, 320–322
M-Audio
    Axiom 49 keyboard, 31
    Oxygen8 v2 keyboard, 30, 31
Media area in Arrange window, 17–18
Media loss parameter, 226
Metronome, 66
Microphones, 35–36
    AKG C1000, 37
    condenser, 35
    connecting of, 36
    dynamic, 36, 37
    electret condenser, 36, 37
    Neumann, 36
MIDI, 46–48
    and audio editing, 7
    channel strips, 267–268
    controller, 233, 234
    controller data
        in Hyper Editor, 190–191
        using Hyper Draw, 180–181
    data, 157, 158, 169
        filtering, 195
    editing,169–170
    files, preparation of, 353–354
    interface, 30
    recording, 11–12, 145, 159

    in cycle mode, 158–160
    overdubs, 158–160
    procedures in, 158–160
    regions, 146
        editing and arranging, 161–162
        fades and crossfades, 161
        merging, 161
        timestretching, 161–162
    sequence, 164, 189
    sockets
        IN, 47
        OUT, 47
        THRU, 48
    sound sources, monitoring, 155–157
    tracks, 157–158
        creating, 146
MIDI Thru function, 166
MIDI time code (MTC), 348
    synchronization settings of,
        351, 352
Mixer, 14, 15, 268–272, 396
    All mode, 270
    Arrange mode, 269, 270
    dropping out sections of, 272
    folders and, 275
    layer, 385
    Single mode, 272
    view modes, 269
Mixing, 259, 275–279
    adding plug-ins, 278–279
    automation. *See* Automation
    aux channel strips application in,
        283–285
    Channel EQ, 280–281
    compression, 276–278, 287
    distortion effects in, 261–263
    equalization, 276–278
    group setting for, 285–287
    insert effects, 280
    part leveling, 275–276
    send effects, 280–283
    vocal processing in, 297–298
Modulation
    EFM1 synthesizers, 220–222
    EXS24 instruments on, 252–253
    sculpture, 232–235
Modulation matrix in ES2, 204–205,
        213–215
Monitor mix, 78
Monitoring, latency and, 51–52
Monitors, 36–38
Monophonic mode, 109
Morphing in sculpture, 234, 235, 236

Tracks (*Continued*)
 parameters, 21
 sorting, 71
 working with, 58–60
Transient markers, 111–114
 audio regions separating based on,
  117–118
Transmit MIDI clock, 351
Transport bar, 63, 64
 in Arrange window, 21
Transposition parameters, 151, 164
Trigger option, 249

U
Ultrabeat, 254–257
 multioutput versions of, 188
Unison mode, 207, 209
Universal audio's UAD system, 46
Universal serial bus (USB)
 devices, 27–32
  audio interfaces, 28–30
  controller keyboards, 30–32
 versions, 28, 29

V
Vector envelope, 235, 236
Vector synthesis, 214–215
Velocity parameters, 151, 164
Velocity range, 245
Video output menu, 336
Video preferences, 338
Virtual instruments, 146–152
 ES E, 165
 ES M, 165

ES P, 166
EVB3, 154–155
EVD6, 160–161
EVP88, 149–150
EXS24. *See* EXS24
Vocal processing, 297–298

W
Wave format file, 365
WaveBurner, 311, 312
 creating fades in, 326
 editing and assembling
  CD in, 328–330
 mastering in, 320–322
Waveshapers, 229–232
Waveshapes, 203–204
Wavetable synthesis, 221
Windows, 377–382,
  384–386
Write, automation mode, 292

Z
Zone menu, 242, 244
Zone parameters
 key range, 244
 loop, 247–248
 output, 246
 pitch, 244–245
 playback, 246
 sample, 246–247
 velocity range, 245
Zoom functions in Arrange
  window, 23–25